The Marketing and Engineering of the Human Body

- Scientists announce the first cloning of human embryos, while the ongoing cloning of sheep and cows creates unexpected "monsters."

- Individuals sell their rare blood to U.S. pharmaceutical companies for up to $6,000 a pint.

- International trade in human organs has reached alarming proportions. Kidneys sell for $10,000 to $40,000.

- Unregulated fetal tissue brokers in the U.S. reap close to a million dollars a year in fetal organ sales.

- Researchers have successfully transplanted fetal organs into laboratory animals, creating "humanized" mice.

- Donors sell their sperm for $50 a donation. Many have "fathered" hundreds of children.

- Women egg donors in New York City receive $2,000 for their ova.

- Worldwide, thousands of babies are bought and sold through surrogate mother contracts.

- In the U.S., thousands of frozen embryos are in legal limbo as the courts decide whether they are people or property.

- Polls show that 11% of American couples would abort a child genetically predisposed to obesity.

- The U.S. National Institutes of Health has applied for patents on over 2,000 human brain genes.

- American companies have attempted to patent women genetically engineered to produce valuable biochemicals in their mammary glands.

- Human genes have been genetically engineered into the permanent genetic code of mice, sheep, pigs, cows, and fish.

- Scientists engineer the AIDS virus into the genetic code of mice, inadvertently creating the possibility of a "super" AIDS which could be transmitted through the air.

The
Human
Body
Shop

The Human Body Shop

THE ENGINEERING AND MARKETING OF LIFE

Andrew Kimbrell

HarperSanFrancisco

A Division of HarperCollins*Publishers*

For my children
Kaiulani and Nicholas

FIRST HARPERCOLLINS PAPERBACK EDITION
PUBLISHED IN 1994

ISBN 0–06–250619–6 (pbk)

**An Earlier Edition of This Book Was Cataloged
As Follows:**

Kimbrell, Andrew.
    The human body shop : the engineering
and marketing of life / Andrew Kimbrell. —
1st ed.
    p.   cm.
    Includes bibliographical references and index.
    ISBN 0-06-250524-6 (cloth)
    1. Biotechnology—Moral and ethical aspects.
    I. Title.
TP248.2.K56   1993
174.25—dc20   90-55774

---

94 95 96 97 98 RRD(H) 10 9 8 7 6 5 4 3 2 1

This edition is printed on acid-free paper that meets
the American National Standards Institute Z39.48
Standard.

CONTENTS

Foreword

IT WAS A far different America in 1932 when Aldous Huxley published his dystopian vision of a *Brave New World*. The book shocked a generation of readers whose tastes and sensibilities were still caught between a lingering nineteenth-century agrarian society and a bustling, new urban industrial order.

Huxley's picture of a future world in which human life is mass produced by precise engineering standards and in accordance with strict utilitarian values struck an emotional chord for Americans wrestling with the promises and perils of modernity. The book both excited and frightened its readers. Its portrait of denatured future beings—biological automatons whose spiritual essence has been severed by science—served as a prescient warning of what might lie ahead on the track of modernity. Still, for many in the intellectual community and the public at large, *Brave New World* was greeted more as a metaphor, a compelling parody of the modern condition, than as a very real possibility.

Today, Huxley's vision is fast becoming commonplace. Engineering principles and mass production techniques are rushing headlong into the interior regions of the biological kingdom, invading the once sacred texts of life. The genetic code has been broken and scientists are rearranging the very blueprints of life. They are inserting, deleting, recombining, editing, and programming genetic sequences within and between species, laying the foundation for a second Creation—an artificial evolution designed with market forces and commercial objectives in mind.

Researchers at the government's National Institutes of Health are currently mapping the entire set of genetic blueprints for the human race, a multibillion dollar, decade-long effort rivaling the government's earlier campaign to land a man on the moon. Anxious to profit from the newest advances in molecular biology, scientists are already staking out patent claims on hundreds of human genes, and by the turn of the century, market analysts expect that the entire human genome will have been patented and made the commercial property of pharmaceutical, chemical, and biotech companies.

The technology of the human body shop is already reshaping our concepts of life. There are now more than a dozen new ways to make a baby, all of which involve sophisticated technological intervention into the reproductive process. The first somatic gene surgery has already been performed on human beings, and scientists predict that the first germline surgery to correct genetic "errors" in eggs, sperm, and embryos will be attempted well before the end of the current decade. The biotech industry looks forward to the day in the not-too-distant future when the entire reproduction process—from conception to birth—will be brought under the watchful eyes of technicians and be made both an efficient production technology and lucrative commercial enterprise.

Where once these technological possibilities sparked awe and a sense of trepidation, they now elicit a mere curiosity. The question, How could we? has been replaced with the questions, How soon? and, At what cost? We are becoming dulled to the new miracles of molecular biology even before the products and processes become available in the marketplace.

In a world where the ideology of the market has triumphed at every turn over traditional values and longstanding beliefs about the sacredness of life and the intrinsic worth of the Creation, it comes as no surprise that the new advances in genetic engineering are being greeted more with open arms than with mass resistance. The technologization and commercialization of the human body marks the final stage of a five-hundred-year journey to enclose, privatize, and commodify the Earth's many ecological commons. We have enclosed the great land masses of the planet, turning whole ecosystems into commercial property. We have similarly appropriated the great oceanic commons. Even the atmosphere has been enclosed into commercial air corridors that can be bought, sold, or leased in the open marketplace. Now we are enclosing and commodifying the human body itself, the last ecological invasion left to the market forces of modernity.

In *The Human Body Shop* Andrew Kimbrell applies his considerable skills as a lawyer and philosopher to the task of dissecting the final stages of the modern age. Kimbrell rightly identifies the commercialization of the human body as the culminating feat of homo economicus. In devastating detail he maps out the market's steady incursion into the human body, pinpointing each new technological accomplishment and accompanying commercial claim. Today, global corporations are swarming over the human body, expropriating every available organ, tissue, and gene.

The colonization of the human body represents a tour de force in the history of modern capitalism, and the last chapter in the desacralization of the human spirit. Computing the price, Mr. Kimbrell warns of the devastating cost to the human psyche of sectioning off the human body and laying it bare on the commercial auction block.

In many respects, Kimbrell's analysis completes an intellectual autopsy begun more than a hundred years ago when an earlier social prophet, Karl Marx, painstakingly traced the historical process that led to the selling of one's time in the form of hourly wages. Kimbrell takes us through the final stages of the despiritualization of modern man and woman with his account of the sale of one's very being in the marketplace.

Kimbrell's work is not likely to shock as much as sadden. It is a painful exegesis of the plight of modern man and woman, increasingly imprisoned by the twin gods of technology and profit. It is a doleful tale of Faustian bargains wrapped in commercial paper, of promissory notes extolling the benefits of a coming technological utopia. The price of entering into this radical adventure has been high.

We have traded away our very souls, says Kimbrell, for the going price of our own parts in the global marketplace.

It is clear upon reading *The Human Body Shop* that the devil is already at the door, cleverly disguised as an engineer and entrepreneur. Mere resistance will not be near enough to fend off this modern day Faust. What is required, argues the author, is a transformation of consciousness, a new understanding of our being and our rightful place and role in the world. Kimbrell asks us to make a leap in faith, to transcend our narrow world of material forces and technological wizardry and renew our covenant with the Creator and the Creation, to act as stewards of our own bodies and beings, and to serve as caretakers of the life spirit of the ages. This is the mission that cries out for our attention on each and every page of Andrew Kimbrell's brilliant requiem for modernity.

Mr. Kimbrell has exposed the shadow side of modern existence to the full light of public scrutiny. It is now up to us to perform the exorcism, to free ourselves from the grip of this fast approaching Brave New World.

Jeremy Rifkin
January 6, 1993
Washington, D.C.

Introduction

ONLY A FEW years ago, the biological materials of the human body had little or no value. Now, a whirlwind of advances in various biological technologies has created a boom market in the body. Organ and fetal transplantation, reproductive technologies, and genetic manipulation have made body parts—even small amounts of our tissues—extremely valuable. Trade in human parts is rapidly becoming a worldwide industry.

In response to the new industry's demand for body materials, more and more people are selling themselves. Donors are selling their blood, organs, and reproductive elements. Researchers and corporations are marketing, and in some instances patenting, human "products," including cells and genes. Body part marketing and sales have allowed a new breed of high-tech entrepreneurs to reap billions of dollars in profits.

This book will take you on a tour of the human body shop. We will start with the sale and manipulation of blood, organs, and fetal parts; proceed to the marketing of human reproductive materials—semen, eggs, embryos, and children; and end with the new biotechnology business of selling and engineering human biochemicals, genes, and cells. We will look into the laboratories, and into the corporate boardrooms, of the designers and marketers of life. We will attend the meetings of the bioregulators at the federal agencies. We will travel the halls of the U.S. Congress and visit courtrooms around the world, to witness the celebrated legislative and legal battles that surround the body shop. We will examine the biorevolution both in the United States and elsewhere, probing the motivations and objectives both of the people who currently run the body shop and its opponents. We will meet the beneficiaries of the marketing and engineering of humans, and the body shop's many victims.

Throughout, we will witness the dazzling short-term benefits offered by the current revolution in the engineering and marketing of life, but we will also see that it raises some of the most profound questions our society has ever had to answer: What is life? What does it mean to be human? Should we allow our scientists to become the genetic coauthors of evolution? How do we define death? Who determines what life is worth living? Do we want a "free market" in human organs, tissues, genes—or children?

The answers to these questions are no longer self-evident. Due to the engineering and marketing of life, the human body, once held to be "sacred" and inviolable, is rapidly becoming the raw material for the new bioindustrial age. The body has become a commodity.

We are currently totally unprepared to cope with the myriad ethical, economic, and political consequences of the human body shop. Clearly, the questions raised by the manipulation and marketing of life are among the most important ever to face the human family, yet we have done little to establish adequate biopolicies to guide us through the moral morass. Policymakers in the forefront of the biorevolution—often goaded by scientific curiosity, the altruistic drive to cure disease, or raw personal profit—do not seem to have the breadth of vision to confront the implications of what they are bringing to society.

Even many of those employed as so-called "bioethicists" appear incapable of saying no to any new advance in the manipulation and sale of life. They seem intent on guiding the unthinkable on its passage to becoming debatable, then justifiable, and finally routine.

The public is told by the experts that the consequences of the human body shop will be profound, but they can be managed. We are led to believe that science will ultimately allow us full knowledge of our bodies, that technology will allow us to have power to cure diseases and perhaps conquer death itself, and that the free market will allow each of us to purchase anything, including life itself. As a society we do not know whether or not to believe the high-tech advertisers. Are we entering a biological utopia, a frightening "Brave New World," or some destination in between? Even if we become convinced that some limits must be put on the manipulation and marketing of life, we feel ill-equipped to judge the complexities of new advances in medicine and genetic technology.

Most remain poorly informed about the biorevolution. In a recent congressional study on the public perception of biotechnology, less than 20 percent of Americans had heard about any dangers posed by biotechnology. This figure is assumed to be typical of most societies in the West where this technology has taken hold. Even those familiar with impacts of designing and selling life have found the bioethical implications of the technology complex and at times overwhelming. Yet even as the public and policymakers remain confused and uninformed, bioengineers at the service of a variety of institutions and corporations are busy at work in thousands of laboratories across the county, spending tens of billions of dollars in pursuit of the mastery and sale of life.

This book attempts to lift the veil of secrecy and ignorance that has kept the human body shop from full public review and comprehension. It will conclude with a number of biopolicies designed to steer our society away from the numerous ethical precipices to which the engineering and selling of life have brought us. These policies reflect not only the current cutting-edge technologies and marketing schemes of the bioengineers, but also the historical and philosophic roots of the manipulation and marketing of the human body. We will see that benefits and risks of various biotechnologies mirror the results of other technologies of our time. Certain body shop technologies, including organ transplantation, offer cures for thousands and hope for many more. The life-saving or beneficial uses of other biotechnologies, including genetic engineering and reproductive technologies, are more uncertain. But those looking to assess the ultimate impacts of the body shop must not forget, or be allowed to forget, that we have never limited a technology—or even approached limiting a technology—to its beneficial uses. Society's past experience with both the nuclear and petrochemical revolution demonstrates that the more powerful a technology is at expropriating and controlling the forces of nature, the greater the disruption of our society and the potential destruction of life as we know it.

Blood and Flesh

1

Blood Tithes

BLOOD IS THE ultimate symbol of life—a magical fluid associated with birth, death, fortitude, nobility, purity, and fertility. Human beings, from time immemorial, have viewed blood as miraculous, have been terrified at the sight of it, and have died rather than betray it. Virtually all cultures have used and treated blood in ways that reflect its symbolic importance. The ancient Egyptians, for example, were said to bathe in blood in order to regain powers lost through illness or injury. They also anointed heads with blood and oil as a treatment for baldness or graying.[2] Blood also plays a central role in the Christian sacramental traditions, where it is transformed into wine and consumed as the essence of life, the very presence of the "living God." As Christ asked of his apostles: "Drink ye all of it; For this is my blood of the new testament. . . ."[3] The sacred symbolic perception of blood also led some cultures to strictly limit its use and consumption. The Old Testament prohibition on eating the blood of animals is based on the belief that blood, unlike flesh, is life itself and should never be consumed. As commanded in Deuteronomy 12:23: "Only be sure that thou eat not the blood; for the blood *is* the life, and thou mayest not eat the life with the flesh."

The life stream of blood runs through each of us regardless of race, creed, or heritage. In many traditions individuals exchange blood as a sign that they have become "brothers" or "sisters." We also look at all warm-blooded creatures with a sense of kinship that transcends species boundaries. Paradoxically, beliefs about blood have torn us apart. Throughout human history, blood has been used as a rationale for persecution and genocide; tribal and racial prejudice have thrived on concepts of "pure" and "impure" blood.

The prohibitions, prejudices, and sacramental rituals that have defined our relationship to blood for much of human history are in stark contrast to its treatment in the twentieth century. Advances in modern medicine, and especially transfusion technology, have caused blood to lose much of its sacred and symbolic meaning. Instead it is often treated as little more than an increasingly valuable commodity in the medical marketplace. Currently, blood is the most commonly sold human body "product."

The open sale of blood created one of the earliest and most bitter controversies over the ethics of establishing a commercial market for human body materials. This conflict has not subsided. Although millions of liters of blood are sold around the world, blood is also the part of ourselves that we donate most often as a gift to the community. As a result of our ambiguous societal stance on blood, its use remains at the very center of the debate over whether human body material is gift or merchandise.

Another Person's Poison

The history of blood as a commodity begins with its value in saving lives through transfusion. According to some historians, the first formal attempt to transfuse blood from one individual to another occurred over five hundred years ago, when an ailing Pope Innocent III received donor blood from three healthy boys. History does not record whether the transfusion was attempted orally or intravenously. Regardless, the result was a disaster: The boys died, the Pope died, and the overly ambitious physician fled the country.[4]

Less than two centuries later, the science of blood took a great leap forward with the publication of a poorly printed 72-page quarto that appeared in 1628 in Frankfurt, Germany. This small pamphlet, a lecture by physician William Harvey called *An Anatomical Dissertation on the Movement of the Heart and Blood in Animals*, transformed the entire Western scientific outlook on blood and the human body. In his historic tract, Harvey redefined the purpose of the heart and clarified the nature of circulation:

> All things, both argument and ocular demonstration, thus confirm that the blood passes through lungs and heart by the force of ventricles, and is driven thence and sent to all parts of the body. There it makes its way into the veins and the pores. It flows by veins everywhere from the circumference to the center. . . . It is therefore necessary to conclude that the blood in animals is impelled in a circle, and is in a state of ceaseless movement; that this is the act or function of the heart, which it performs by means of its pulse; and that it is the sole and only end of movement and pulse of the heart.[5]

Harvey went on to compare the heart and circulation with the mechanical "water pumps" in use in his time.[6]

In 1665, buoyed by Harvey's revelations on the circulation of blood, English architect Sir Christopher Wren goaded noted anatomist Richard Lower into performing a historic transfusion. Lower accomplished the blood transfer by uniting the artery of one dog to the vein of another with a hollow quill. The operation was a success. Lower's experiment caused a considerable stir among the leading lights of the early English Enlightenment. Robert Boyle, known as the father of chemistry, tried the transfusion experiment himself. After ascertaining its efficacy, Boyle lost no time in enthusiastically recommending the use of a human guinea pig for the experiment: "Tryal might be made upon some humane bodies, especially those of Malefactors."[7]

Boyle got his wish in November 1667. Lower, in a demonstration for the Royal Society in London, transfused a healthy but mildly insane man with the blood of a lamb. The man got twenty shillings for his trouble. A week later diarist Samuel Pepys, who would himself become president of the Royal Society, visited the adult transfusion patient and found him "much better."[8] Actually, the first transfusion of animal blood into a human had occurred several months before, when Frenchman Jean-Baptiste Denis transfused the blood of a boy of fifteen who had been suffering from a persistent fever for months. The unfortunate patient had been bled over twenty times in the attempt to cure him. Denis gave the boy half a pint of lamb's blood and he apparently recovered, although it is difficult to ascertain whether the recovery was due to the transfusion or the cessation of bloodletting.

Even as primitive techniques of transfusion were developed, many myths about blood remained. For example, people still believed that many individual or species traits were "carried" in the blood. As transfusions became the rage in the late seventeenth century, many speculated that blood transfusions would also create an exchange of traits. Boyle questioned whether frequent transfusions of mastiff blood into bloodhounds might "prejudice them in point of scent."[9] Pepys came up with several original and humorous suggestions, including causing the "blood of a Quaker to be let into an archbishop" and awaiting the theological result.[10]

Despite early success Denis and his transfusion cohorts were headed for disaster. They did not know, of course, that animal blood is not compatible with human blood. Denis was greatly surprised when the next patient he tried to transfuse with animal blood died. After the fatal experiment, the widow of the deceased patient charged Denis with murder and initiated a successful legal battle against him. The Denis trial confirmed the suspicion of many that blood was far too mysterious and powerful a substance to be used in transfusions. Over the next few years, the bad publicity resulting from the trial, and other transfusion-related deaths, caused the French, Italian, and English governments, and the Pope to ban all transfusions.[11]

Even with the ban in place, and transfusions rare, new insights into the functioning of blood continued. In 1777, the French chemist Antoine-Laurent Lavoisier, who had become aware that combustion required oxygen (which he named), was also alerted to the idea that physiological respiration is a similar process: that the body takes in oxygen through the lungs and into the blood, where it is circulated through the body then and slowly "burned." Lavoisier was ill-rewarded for his remarkable discovery. In 1794, he was beheaded, as the new moguls of the French Revolution exclaimed that they needed no savants. Upon his execution a friend was to remark, "Only a moment was required to sever that head, and perhaps a century will not be sufficient to produce another like it."[12] It didn't take quite that long, however. In 1818, James Blundell, an English obstetrician, conducted several transfusion experiments that cast light on the failures of Denis and others. He demonstrated that dog blood was not compatible with the blood of other mammals and was therefore likely to be harmful to humans. Blundell is generally credited with being the first to perform a human-to-human transfusion.[13]

The great breakthrough in blood transfusion took place at the dawn of the twentieth century. In 1900, Karl Landsteiner, a young Austrian researcher, discovered the first blood groups. He detected the presence of chemical substances on red blood cells, which can react with the red cells in certain other human subjects with potentially fatal results. These chemical substances came to be called A and B. According to whether people had one or the other, or both, or neither, four blood groups were determined and named A, B, AB, and O. After Landsteiner it became clear that if blood were to be taken from one human and transfused into another, "incompatible" red cells should not be used. The wrong blood type would cause a potentially fatal antibody reaction. He realized that one person's blood was another person's poison.[14]

During the first half of the twentieth century, the tragic toll of two world wars in countless broken and destroyed bodies necessitated massive increases in blood transfusions. Battlefield transfusions were given in the tens of thousands; ten to twenty donor transfusions were routine for seriously wounded soldiers. With this volume of cases, surgeons gained significant data on transfusions, including observation of a number of reactions between supposedly compatible blood types. Additional research based on these observations resulted in further subdivision of the four blood groups. Another milestone was the development of the Rhesus system of blood analysis in 1939 and 1940 by Landsteiner (who was the leading researcher in this field for over four decades): eventually, eight blood groups were distinguished.[15]

Toward the end of World War II, as knowledge about blood increased and transfusion techniques advanced, the use of human blood became an increasingly vital factor in U.S. medical care. As transfusions began to be performed more frequently, the federal government stepped in to regulate the new industry. In 1944, Congress required that blood banks obtain licenses, and established federal supervision and control over the inspection of blood banks and the issuance of licenses permitting the transportation of blood in interstate commerce. In 1947 the American Association of Blood Banks (AABB) was formed and would become the main national organization of U.S. blood banks. By the late 1950s, over 5 million pints of blood were used for transfusions each year.[16]

Almost three centuries after the first scientific attempt at blood transfusion, the technology had come of age. Innumerable lives were being saved, and there was an unprecedented boom in blood value and demand. At this critical juncture, a historic struggle began over how blood was to be obtained and supplied. Was blood a "gift," retaining some of its symbolic sacredness, to be donated by a community as an act of social benevolence? Or was blood a new commodity, a product that could now be advertised and sold as any other in the post–World War II economic boom? The controversy about blood that developed in the 1950s and 1960s became the greatest public debate about the ownership of human tissues since slavery. As economist Richard M. Titmuss remarked over twenty years ago, "Short of examining humankind itself and the institution of slavery—of men and women as market commodities—blood as a living tissue may now constitute in Western societies one of the ultimate tests of where the 'social' begins and the 'economic' ends."[17]

Restraint of "Vampires"

The first great public battle over the human body shop began with a legal action initiated in 1962. The historic case, *Community Blood Bank of Kansas City, Inc. v. the Federal Trade Commission*, was the first to raise the question of whether blood or other body parts could be defined as commodities, no different than any other product.

The controversy that led to the landmark case began in Kansas City, Missouri, in the early 1950s. As elsewhere around the country, Kansas City had an increased demand for blood but inadequate supply. In 1955, a commercial blood bank, which paid its donors, was set up to meet the growing blood demand. The commercial bank was called the Midwest Blood Bank and Plasma Center. The Midwest Bank was not a particularly savory operation. The building was situated in a slum area and displayed a sign reading "Cash Paid for Blood." Many of the bank's paid donors were later described in federal hearings as "skid-row" types—homeless men and women who had records of being institutionalized, or were chronic alcoholics or drug addicts driven to donation to pay for food or their addiction. Some critics referred to the bank as a "vampire" on these poor individuals.

Reportedly, the bank's operations were bizarre and unsanitary. One witness reported having seen "worms all over the floor." A doctor remembered an incident during which Midwest attempted delivery of blood in a beer can. Moreover, Midwest's administrators did not inspire confidence. The bank was owned and operated by a married couple. The husband had only completed grade school, had no medical training, and had previously worked as a car salesman and banjo teacher. His wife claimed to be a registered nurse but was not licensed in either Kansas or Missouri. The husband had reportedly chased a third partner out of the bank at gunpoint. Nevertheless, Midwest Bank was given its license by federal regulators, and in 1958 its owners even opened up another blood bank, World Blood Bank, Inc.[18]

At about the same time, local citizens, doctors, and hospital administrators joined together to form a nonprofit community blood bank that would supply the community with blood from unpaid volunteers. Despite its good intentions, the blood bank project got off to a poor start. The diverse parties who were attempting to set up the bank simply could not agree on its organizational structure. After several years of negotiations, the Community Blood Bank of the Kansas City Area was finally incorporated in April of 1957 and began operating a year later.[19]

Not surprisingly, given Midwest's generally poor reputation, as soon as the nonprofit Community Bank began operating in 1958, practically all the hospitals, doctors, and pathologists in the Kansas

City area entered into agreements to use its blood. Within several months Community gained a virtual monopoly on supplying blood in Kansas City. The two commercial banks, Midwest and World, were outraged at the quick success of the nonprofit blood bank and suspected that hospitals, doctors, and the Community Bank were colluding to put them out of business. In July 1962, Midwest Bank and World Bank took an extraordinary legal action. They filed a complaint with the Federal Trade Commission (FTC) against the Community Bank and its officers, as well as several hospitals, hospital administrators, and doctors in the Kansas City area, claiming that the alleged culprits had created a "conspiracy" not to use the paid donor blood from the commercial banks and had attempted to get others not to use the blood. As stated in the complaint, "[Defendants] entered into and have since carried out an agreement, understanding, combination or planned course of action . . . to hamper, restrict and restrain the sale and distribution of blood in interstate commerce."[20] The complaint was the first ever to charge restraint of trade in blood or any other human body element.

In September 1963, a lengthy series of hearings were held before FTC examiner Walter K. Bennett. Several months later Bennett held that Community Bank and the other accused were guilty as charged. He concluded that the defendants had conspired to restrain trade in the "commodity" of blood. The decision shocked and dismayed many in the medical community. Doctors were concerned that the decision would mean that they could not freely choose to use nonpaid blood in their practices lest they be sued by a commercial blood bank for restraint of trade. Community Bank immediately appealed, and the case went to the full five-member Federal Trade Commission for a decision.

Throughout the case Community Bank and the others accused maintained that the blood was not a commodity for sale. Blood, they argued, was living human tissue used in a medical service; it was not a "product" like a car, or even a drug. They maintained that blood could not and should not be part of commerce, and as such the Federal Trade Commission had no power to regulate its use or penalize those who tried to prevent the sale of paid blood. They insisted that the decision to use paid or nonpaid blood was moral, not economic.

Outside associations, including the American Medical Association, the Red Cross, and the AFL-CIO, submitted statements maintaining that "trafficking in blood" was unethical and immoral.[21] Doctor after doctor testified that they would not use blood from paid donors, because the "purchase and sale of human blood, or any other part of the human body, is wrong." They also noted that blood could not be a "product" because it is wholly made by the human body as part of its own functioning; it was not manufactured for sale and consumption. They argued that calling blood a commodity was a dangerous "fiction" that would allow the market to be extended into other living parts of the human body. If you start with blood as a commodity, where will it end? they asked. If blood is a commodity, how much of the rest of the human body is for sale as a product?[22]

As the hearings proceeded, the controversy, which had been getting more and more public attention, prompted action in Congress. During 1964 and 1965, Senator Edward V. Long of Missouri struggled to pass legislation that would exempt nonprofit blood banks from antitrust legislation and officially categorize blood as a noncommodity. The legislation received widespread support, but it was not passed.[23]

In September 1966, with congressional action at a standstill, the FTC ruled on the case. By a three-to-two margin, the commission's opinion upheld the original examiner's decision that the Community Blood Bank, and the hospital, doctors, and others charged with them, were guilty of conspiring to restrain Midwest Bank's and World Bank's trade in blood. The FTC noted in no uncertain terms that "whole [human] blood may, for purposes of Section 5 of the Federal Trade Commission Act, be considered to be a 'product' or a 'commodity.'"[24] The opinion continued,

> There is a sufficient basis in the record for a factual conclusion that whole [human] blood is a "biological product." . . . As a result the commercial banks in this case when acquiring, processing, and supplying such blood to hospitals are engaged in the business of producing and selling a product. . . . The Commission clearly has jurisdiction to proceed against a conspiracy designed to have the effect of hindering the operation of such a business, and we so hold.[25]

Two of the five commissioners vigorously dissented from the majority decision.

In 1967, aggrieved by the FTC decision, Senator Long once again introduced legislation in the Senate to declare blood a noncommodity and blood banks free of monopoly regulation. Once again it failed to pass. With the FTC opinion and Long's second failure in Congress, it appeared as if there had been a complete victory for those advocating the commercialization of blood.[26]

However, Community Bank had appealed the FTC ruling to the Eighth Circuit Court of Appeals. In January 1969, the circuit court dramatically reversed the FTC decision. The court declared that certain nonprofit corporations, including blood banks like Community, were immune from FTC regulation. This allowed the medical profession free rein to exclusively use nonprofits without running afoul of antitrust or restraint of trade laws. Though the court appeared sympathetic to the idea that blood was not a commodity for the purpose of FTC regulation, it did not rule specifically on the issue.[27]

In the years following the *Kansas City* case, the struggle over the status of blood intensified. New public outrage at the commercialization of blood was stirred by several grisly reports of "vampirism" in Third World countries. These reports cited numerous instances where highly profitable blood banks had extracted plasma from undernourished and economically disenfranchised donors around the world, and then sold the blood at high prices.

One of the most egregious instances involved Anastasio Somoza, the brutal dictator whose family occupied the Nicaraguan presidency for nearly half a century. In 1973, Somoza and a Cuban exile, Dr. Pedro Ramos, opened a plasma center called "Plasmaferesis" in Managua. The center reportedly bought blood from the "poor and undernourished" and also forced political prisoners to donate blood. The facility was licensed by the U.S. Food and Drug Administration and the plasma collected was sold primarily to the United States and Western Europe. Each year between 1973 and 1977 up to 300,000 "donations" were collected, two-thirds of which were sold for export.[28]

The center was virtually unregulated and reports indicated that numerous donors desperate for money died due to too-frequent donations. One long-time donor at the center described his motivations for selling blood:

You see poverty forces me to. I need money and that means that I have to come here often. I have a brother who is very ill and he has not a job, and I have only a little work now and then and that doesn't give me enough when I have a big family. And it is hard to get work here in Nicaragua. And we are all so poor, you see. . . . I am cold. I am shaking. . . . You feel weak all the time, you get weaker and weaker every time you come here and give blood. . . . I've gone down a lot in weight. It feels as though everything disappears in you. . . . Here they are only interested in seeing that people come and give blood. They don't care about us and how we live. It is hard to believe that they are doctors. . . . You can just go and die in the gutter, nobody pays attention or cares. All they are interested in is taking our blood.[29]

Certain courageous members of the press began exposing conditions at the blood-selling facility. Pedro Joaquin Chamorro, editor-in-chief of Nicaragua's leading opposition paper, *La Prensa*, described Somoza's activities as an "inhuman trade in the blood of Nicaraguans." The National Red Cross, church officials, and national medical associations supported Chamorro's reports. Angered and frightened by the growing criticism of Somoza's regime, allies of the dictator hired assassins, who shot down the editor on an open street in January 1978. One of the killers later admitted to receiving over $14,000 for committing the murder.

During Chamorro's funeral thousands of Nicaraguans gathered to protest. They chanted, "Who killed Chamorro? Somoza!" The outraged crowd then rampaged through the city and, accompanied by cries of "Vampire Somoza!" burned down the plasma center. The demonstration signaled the beginning of the end for Somoza. Eighteen months later he had fled to Paraguay. Pedro Ramos had already fled to the United States, where he was to reestablish his blood business in Miami. Over a decade later, Chamorro's wife, Violeta, was elected as president of Nicaragua.[30]

By the early 1970s, alarm over the increasing worldwide sale of blood prompted various international agencies to act. In 1973, the League of Red Cross Societies issued a warning about the rapidly expanding blood market, and in 1975, the 28th World Assembly passed a resolution calling for nonpaid donor systems to be set up in all member countries. Even the United States encouraged volunteerism when in 1972 President Richard M. Nixon announced a National Blood policy to encourage nonpaid blood donations.[31]

The battle against commercialization also received a boost from the publication of *The Gift Relationship: From Human Blood to Social Policy*, by English author and economist Richard Titmuss. Published in 1971, Titmuss's immensely influential book urged countries to move toward wholly nonpaid donor systems blood collection. He argued for the use of nonpaid donors, primarily for the moral reason that such donation made individuals and their communities more generative. He noted that each act of donation was not only a gift, but also an act of faith that others would give.

> In not asking for or expecting any payment of money these [blood] donors signified their belief in the willingness of other men to act altruistically in the future, and to combine together to make a gift freely available should they have a need for it. By expressing confidence in the behavior of future unknown strangers they were thus denying the Hobbesian thesis that men are devoid of any distinctively moral sense. . . . As individuals they were, it may be said, taking part in the creation of a greater good transcending the good of self-love. To "love" themselves they recognized the need to "love" strangers. By contrast one of the functions of the atomistic private market systems is to "free" men from any sense of obligation to or for other men. . . .[32]

The "gift concept" of blood donation slowly advanced over the next two decades, and not only for altruistic reasons. Volunteerism was often supported as an aid in insuring a safe blood supply. From the beginning people understood that blood transfusions are an extremely efficient way to pass on disease. It was believed that paid donors, who generally came from the poorer segments of the population, were more likely to be infected with bacteria, such as those causing syphilis, or viruses such as hepatitis. Vigilance in screening blood for transfusions began as early as 1947, with the initiation of a test to detect syphilis. In the 1950s and 1960s, the blood supply was seriously undermined through contamination by the hepatitis B virus, and studies showed that numerous transfusion patients were being infected with hepatitis. These studies generally supported the view that non-paid donor blood was less infected with the dangerous virus than was paid blood. In 1971, a screening test was devised to detect blood infected with hepatitis B. However, according to the American Red Cross, the risk of coming down with symptoms of hepatitis following a blood transfusion is still about one case per thousand.

What hepatitis was for the 1960s and 1970s, AIDS became for the 1980s. In the early 1980s, the world was shocked by discovery of the AIDS virus (HIV) and its widespread infection of victims on several continents. Despite immediate concern for the blood supply, it took until 1985 for researchers to find a reliable HIV screening process for donated blood. For many it was too late. Exact numbers of those infected through transfusion is not known, but certainly hundreds were infected each year prior to the installation of the screening process. The Centers for Disease Control now estimates that the average transfusion patient has about a one in 75,000 chance of receiving HIV-contaminated blood. Currently, blood is also being screened for HTLV-1, hepatitis C, and HIV-2.[33]

For a combination of ethical and practical reasons, blood volunteerism finally triumphed in the United States. In the twenty-five years following the 1966 decision in *Kansas City*, the use of paid donors for whole blood used in transfusions has declined in the United States from about 80 percent of the blood market to less than 1 percent. In the process, giving blood has become an important way for hundred of thousands of Americans to express compassion for their fellow men and women and devotion to their community. Each year about 13 million volunteer donations are made, and the shortages predicted by those recommending a free market in blood have not materialized. Since 1984 there has not been a significant national shortage in blood.[34]

Although the vast majority of blood used for transfusions is now obtained from nonpaid donors, the sale of blood and its exploitative results have not been halted. The "vampires" have not been restrained. Blood is still being bought and sold around the world as a growing part of the pharmaceutical industry. Over the last two decades, blood and blood products have quietly become a major medical commodity. A variety of antibodies, clotting factors, and other blood products have become a gold mine for national and international pharmaceutical companies. Increasingly, companies and brokers buy blood from donors in order to fractionate it for use in the production of vaccines, diagnostics, and other drugs. Some donors have found that their blood is worth hundreds, even thousands of dollars per pint.

The "OPEC" of Blood

The following advertisement is run regularly in a San Diego weekly newspaper:

MEASLES □ LUPUS □ CHICKEN POX □ HEPATITIS
MUMPS □ HERPES □ RH SENSITIZED □ MONO

YOUR
ILLNESS
COULD BE
PROFITABLE
$$$

Do you have lupus or have you recently recovered from
a major illness? If so, your plasma could make a valuable
contribution to the medical industry and earn you
$$ at the same time[35]

M.D. Laboratories, the company sponsoring the ad, is one of hundreds of commercial blood centers and pharmaceutical laboratories that pays "donors" to sit for a "bleed"—the removal of a pint of blood, the extraction of the valuable plasma, and the reinjection of the remaining red blood cells into the body. The process is called *plasmapheresis;* and because most of the blood is returned to the body, donors can sit for as many as two bleeds a week.

About 55 percent of blood consists of plasma, a fluid that is composed of 92 percent water, 7 percent proteins, and 1 percent other substances. Early in World War II, researchers discovered efficient techniques for separating plasma from other blood components and for fractionating plasma itself in order to use plasma components in treating patients. Techniques for fractionating plasma have continued to become more sophisticated over the last half century. As noted in the ad, when plasma is separated out, the plasma proteins contain valuable blood products, including important blood-clotting factors and immunoglobulins (antibodies), all useful in treating many diseases and conditions.[36] M.D. Laboratories runs the ad in the hope of finding donors who carry valuable antibodies in their blood—antibodies that can be used for passive vaccination and as diagnostic tools for diseases. To those whose blood contains antibodies for lupus or hepatitis A, M.D. Laboratories pays a fee of between $50 and $200 per bleed. Measles victims can earn as much as $300, and those who lack a blood-clotting factor can make up to $600.[37]

Some rare blood donors do even better. One celebrated instance involved the blood of Ted Slavin. Slavin was a hemophiliac who contracted hepatitis B in the mid-1950s, probably through the numerous blood transfusions his illness forced him to undergo. In 1970 the Hemophilia Society began testing hemophilia patients for hepatitis B antibodies, and Slavin was among those tested. Slavin's test revealed that his blood contained an extraordinarily high concentration of antibodies for the hepatitis B virus. Slavin quickly realized that his test results meant that his blood would be very valuable both to commercial organizations that were using human serum to manufacture diagnostic kits for hepatitis B virus, and to laboratory researchers investigating blood-borne diseases. Slavin began marketing his rare blood to commercial organizations for the equivalent of $6,000 a pint. He also donated blood to noncommercial hepatitis researchers for free. Slavin soon became a blood entrepreneur, forming his own company, Essential Biologicals, which marketed his blood and the blood of others that contained rare factors. Slavin died of hemophilia and other complications in 1984.[38]

Today more than four hundred U.S. commercial blood centers collect, buy, and market blood products. It is estimated that in 1991 13 million plasmapheresis procedures were performed by these commercial blood centers in the United States. Over 95 percent of plasmapheresis donors were paid. These procedures resulted in the purchase of approximately 7 million liters of source plasma for further breakdown and manufacture into various products, such as a variety of antibodies, antihemophilic concentrates, and albumin. Voluntary donor centers like the Red Cross provide another 2 million liters of plasma, which they have collected for free from donors, but which they sell at market price in the plasma products market. Worldwide, 15 million liters of plasma are obtained each year.

By the mid-1980s, the United States had become a leading producer and exporter of plasma "products." Currently, the United States uses only about 22 percent of all plasma products (Europe uses 28 percent and Japan 47 percent), but collects 60 percent of all such products. The U.S. plasma products surplus has allowed it to become the world leader in exporting blood, which internationally is a $2 billion industry.[39] One commentator has called the United States "the OPEC of blood."[40]

My Mother, the Container

The legal and ethical consequences of the increasing commercialization of blood remain uncertain. However, some recent cases, in courts around the country, give a troubling future vision of how the commercialization of blood could affect the legal definition, not only of blood, but of the human body itself. Such a case was the one that the Internal Revenue Service filed against Margaret Cramer Green in 1980.

Green, a woman of modest means and the mother of three teenage children, sold her blood in order to make ends meet. For several years, including 1976, her main source of income was the sale of her rare form of blood, a variant of AB negative, to Serologicals, Inc., a commercial blood laboratory in Pensacola, Florida, twenty miles from her home in Milton. Using plasmapheresis, the company extracted the plasma from Green's blood—two bleeds produce one pint of plasma—and returned the red cells to her body. Plasmapheresis allowed Green to make donations about twice a week, and in 1976 she made ninety-five donations. However, she required a special diet and medication to keep up the quality of her plasma.

In 1976, Ms. Green received over $7,000 in gross receipts from her donor activity—$6,695 in donor commissions and $475 in travel allowances. Green's legal troubles with the IRS started when she claimed $2,355 in business expense deductions to offset her blood income. Green claimed as deductions the loss of certain minerals and antibodies in her blood, transportation to and from the laboratory where she donated blood, medical insurance premiums, special drugs, and the expense of her special diet. The IRS did not agree, and determined that she owed $577 of additional income tax.

Green took the matter to tax court, and the court had a field day with the novel legal questions presented. First the court ruled that blood was a commodity, and that Green was essentially both a "factory" and a "container" of her product, blood. The court held that she was not involved in a service, but rather "the usual sale of a product by a manufacturer to a distributor of raw materials, by a producer to a processor. A tangible product changed hands at a price, paid by the pint."[41] The court noted that other organic products were sold, albeit not "human" products, and that blood was just another such product:

> The rarity of the petitioner's blood made the processing and packaging of her blood plasma a profitable undertaking, just as it is profitable for other entrepreneurs to purchase hen's eggs, bee's honey, cow's milk or sheep's wool for processing and distribution. Although we recognize the traditional sanctity of the human body, we can find no reason to legally distinguish the sale of these raw products of nature from the sale of the petitioner's blood plasma.[42]

Noting the "unique nature of the manufacturing machinery in this instance [Ms. Green]," the court granted her the deductions for the price of a high protein diet as a manufacturing business expense. The court also allowed Green to deduct travel expenses for her ninety-five trips to and from the clinic. The court noted that Ms. Green was not traveling as a "person"—personal travel is not deductible—but rather she was "the container in which her product was transported to the market." You can deduct as a business expense the transportation of containers of a commodity to the place of sale. The court was less forthcoming on Green's attempt to deduct for the lost mineral resources of her blood, finding that Congress intended the deduction of mineral resources to apply to "geological mineral resources" not "the bodies of taxpayers."[43]

Whatever its comic overtones, the Green case underscores the distance we have traveled in our treatment of blood and the human body—from the magical, sacramental definitions of antiquity to the market definitions of today. The former covenants of blood with the divinities have now become commercial contracts with pharmaceutical companies. The prohibitions on the use and consumption of blood have given way to its open sale in the marketplace. For legal purposes even the human body can now be seen as merely the "factory" and "container" of this new valuable biomedical product.

As courts and legislatures struggle over these new definitions of blood and the body, the public debate about the appropriateness of its sale continues. For some the ethical issues surrounding the commodification of blood are not serious. Blood, they say, is replenishable. Its extraction is relatively risk-free. The donor quickly makes more of the product. Yes, the symbolic meaning of blood is reduced, and market forces rather than communitarian altruism have become the basis for certain types of blood collection, but lives are being saved and the exploitation of those selling blood is minimal.

For decades, however, many others have feared that blood was just the beginning of the slippery slope toward the commodification of the body, and history has proved these critics of blood sale correct. For just as transfusion technology and advances in blood science were creating a lucrative market for blood and blood products, transplantation technology and new surgical techniques were making human organs a valuable new commodity. However, unlike blood, organs are not replenishable. Nor is their retrieval risk-free. Moreover, organ harvesting from the dead, the living, and the unborn raises frightening questions about changing the very definitions of life and death. By the early 1980s, the growing commercial market for body parts was creating ethical and economic concerns that the U.S. Congress, and legislatures around the world, could no longer ignore.

2

*And the Lord God caused the man to fall into a deep sleep;
and while he was sleeping, he took one of the man's ribs
and closed up the place with flesh . . . and made a woman
from the rib he had taken out of the man.*
  Genesis 2:21, 22

We are doing a thriving business.
  *Dr. Atma Ram, who sells kidneys taken from patients
  at the Indian Institute of Medical Sciences in New
  Delhi*[1]

Transplanting Profits

THE IDEA OF transplanting human limbs, organs,
bones, and tissues has excited the human imagination throughout
history. Genesis recounts the creation of Eve from Adam's rib. Greek
myths tell of transplants of animal parts to people—most memorably,
the ill-fated construction of prosthetic wings by Daedalus and his son
Icarus. Even the concept of transplanting body parts from cadavers to
patients found its way into the literature and art of antiquity. Among
the most famous medieval portrayals of transplantation is a fifteenth-
century painting attributed to the Italian artist Girolamo da Cre-
mona. The canvas illustrates the legend of Saints Cosmas and
Damian, the patron saints of physicians, transplanting the leg of a
dead donor to a saintly nun, whose cancerous leg is being replaced.
A white-bearded God, accompanied by a choir of angels, serenely
and approvingly surveys the scene.

According to at least one medical historian, early human tissue transplantation was not just myth or miracle. He reports that
"Skin flaps from the forehead and the cheeks were used by Indian
surgeons more than 2,000 years ago, principally in the operation

of rhinoplasty [rebuilding noses]."[2] It was not until the nineteenth century, however, that Western medicine significantly utilized skin transplantation. In 1804, early transplantation pioneers reported successful skin grafts on animals. Toward the end of the century, numerous skin grafts from human to human were also performed. Reports on these procedures are sketchy, and generally consist of firsthand accounts of those involved. A vivid example is Winston Churchill's account of how he donated a piece of skin to a fellow officer injured during the Sudanese war in 1898:

> Molyneux had been rescued from certain slaughter by the hero-ism of one of his troopers. He was now proceeding to England in the charge of a hospital nurse. I decided to keep him company. While we were talking, the doctor came in to dress his wound. It was a horrible gash, and the doctor was anxious that it be skinned over as soon as possible. He said something in a low tone to the nurse, who bared her arm. They retired to a corner, where he began to cut a piece of skin off her to transfer to Molyneux's wound. The poor nurse blanched, and the doctor turned upon me. He was a great raw-boned Irishman. "Oi'll have to take it off you," he said. There was no escape, and as I rolled up my sleeve he added genially, "Ye've heeard of a man being flayed aloive? Well this is what it feels like." He then proceeded to cut a piece of skin and some flesh about the size of a shilling from the inside of my forearm. My sensations as he sawed the razor slowly to and fro fully justified his description of the ordeal. However, I man-aged to hold out until he had cut a beautiful piece of skin with a thin layer of flesh attached to it. This precious fragment was then grafted onto my friend's wound. It remains there to this day. . . .[3]

Organ transplants in humans were not attempted until the first decade of this century. Early ventures included several unsuc-cessful and ethically questionable tries at transplanting kidneys from animals to humans. The major breakthrough in transplantation re-search occurred in the 1940s through the work of Sir Peter Medawar. Medawar, inspired by the findings of fellow researcher Sir Frank Burnet, explained how the body's immune system recognizes and then rejects foreign body materials that enter it. This knowledge led to the development of tissue typing, where a donor's and recipient's tissues are examined in order to attempt compatibility. His discoveries set the stage for modern transplantation. Medawar and Burnet shared the 1970 Nobel Prize for Medicine for their discovery.

The first human-to-human vital organ transplant was performed by Dr. David Hume in 1951. He transplanted the kidney of a cadaver into a failing patient in an unsuccessful attempt to save the patient's life. Over the next four years, Hume, along with Dr. Joseph Murray, performed ten other kidney transplants using cadaver donors. All failed. On February 11, 1953, they achieved a moderate success when one of their patients lived six months with a transplanted kidney. Then in 1954, Murray and his team at the Peter Bent Brigham Hospital in Boston carried out what is generally regarded as the world's first successful kidney transplant. Richard Herrick received a kidney from his identical twin brother, Ronald, and lived eight years before dying of a heart attack. Kidney transplants became increasingly common over the next two decades. By 1975, over 25,000 kidney transplants were carried out around the world, the longest survival being twenty years.[4]

No other organ transplant has been as successful, or as widespread, as kidney transplants. Almost ten years after the first kidney transplant took place, Drs. William Waddell and Thomas E. Starzl at the Veteran's Administration Hospital in Denver, Colorado, performed the world's first liver transplant. The recipient, a forty-seven-year-old janitor named William Grigsby, died within three weeks. By the early 1980s, only a few hundred liver transplants were performed around the world. The longest survival was nearly seven years.

Among the least successful organs for transplant have been lungs. The first lung transplant in history was performed by Dr. James Hardy at the Mississippi Medical Center in June of 1963. The transplant surgery was controversial because the patient, John R. Russell, was a convicted murderer serving a life sentence. There was concern that Russell was simply used as a guinea pig in that he "had nothing to lose." He died eighteen days after the transplants. By the beginning of the 1980s, only about forty lungs had been transplanted, and the longest survival was only ten months.[5]

The transplant breakthrough that seized the public imagination more than any other was the first heart transplant. On December 3,

1967, South African surgeon Dr. Christiaan Barnard transplanted the heart of a young woman car-crash victim into fifty-five-year-old grocer Louis Washinsky, at Groote Schuur Hospital in Capetown, South Africa. After the surgery Washinsky commented, "I am a new Frankenstein." He lived for only eighteen days with his new heart. Barnard's second patient, Dr. Philip Blaiberg, a fifty-eight-year-old dentist, did better; he survived eighty-four weeks. The mystical and symbolic nature of the human heart gave Barnard's surgical feat immediate media allure. The photogenic Barnard, a youthful looking forty-five years of age at the time of the first heart transplant, became an instant international "star." In following years he was frequently featured on the society page hobnobbing with the rich and famous. Barnard's meteoric rise caused a flurry of imitations. By the next year, sixty-four copycat teams had attempted 107 heart transplants in twenty-two countries. However, within months the "gee whiz" atmosphere of the press turned gloomy, as most of the subsequent heart transplants turned into grisly failures. Medical journals began running articles critical of the heart transplant hoopla. Headlines such as "Too Many, Too Soon" became common. Due in good measure to the early irresponsibility surrounding the procedure, the re-emergence of heart transplants as a viable treatment would take almost fifteen years.[6]

Organ transplantation finally came of age in the 1980s. Thanks to better surgical techniques, greater understanding of the body's immune system, and the development of effective drugs to combat rejection, survival rates for those undergoing transplantation improved dramatically. With increased survival came an exponential growth in transplants. Since 1982, the yearly number of heart transplants has increased twenty times, liver transplants forty times, and kidney transplants have doubled. In 1991, a record 16,003 transplants were performed in the United States, including 9,943 kidneys, 2,127 hearts, 2,946 livers, 535 pancreases, 400 lungs, and 52 heart/lungs. Additionally, approximately 30,000 cornea transplants were made.[7]

The transplantation revolution of the last decade has turned the body into a bewildering number of reusable parts. Vital organs, skin, and a variety of other body parts and materials have all become potential commodities in the human body shop. Consider that a donor can deliver:

2 corneas to help restore sight

2 each of the inner ear, the hammer, anvil, and stirrup, to ameliorate some forms of deafness

1 jaw bone used in facial reconstruction

1 heart

1 heart pericardium (the sac that surrounds the heart is made of tough tissue that can be used to cover the brain after surgery)

4 separate heart valves

2 lungs

1 liver

2 kidneys

1 pancreas, which when transplanted can restore insulin production in diabetics

1 stomach (stomachs have been transplanted experimentally without much success)

206 separate bones, including long bones of the arms and legs for use in limb reconstruction and ribs used in spinal fusion and facial repair

2 hip joints

about 27 ligaments and cartilages used in rebuilding ankles, knees, hips, elbows, and shoulder joints

approximately 20 square feet of skin, which can be used as a temporary covering for burn injuries

over 60,000 miles of blood vessels, mostly veins that can be transplanted to reroute blood around blockages

and nearly 90 ounces of bone marrow to treat leukemia and a variety of other diseases.[8]

The new use of the body as spare parts was graphically, if tragically, demonstrated by the case of William Norwood. Norwood, twenty-two, was killed in a 1985 robbery. Physicians quickly made use of his body. Within a short time of his death, his body parts were transplanted into no less than fifty-two different people. Unfortunately, it was later discovered that Norwood had been HIV positive and at least four of the recipients of his body parts contracted or died of AIDS.[9]

As more and more of the body became transplantable, the business of transplantation has skyrocketed. Currently, the cost of transplants runs into several billion dollars each year. The average heart transplant can cost well over $100,000. Liver transplants can run over $200,000, and kidneys average $25,000. Fortunately, in the U.S., federal Medicare and Medicaid programs now cover a significant portion of the costs for the most common transplants, including follow-up drug treatments for up to one year. However, even for the insured, multiyear costs for required drugs can be overwhelming.[10] Additionally, as transplants have become more effective, demand for body organs and parts has far outstripped supply. By the end of 1991, over 23,000 Americans were on an ever-growing waiting list for donated organs. A new name is added to the list every thirty minutes. In 1990, an estimated 2,206 people died waiting for an organ.[11]

With the new and urgent demand for organs, the early 1980s saw the beginning of a historic debate on whether there should be an open market on the body's reusable organs and tissues. Just as conflicts over the status of blood dominated the legal and bioethical debates about the body in the 1960s and 1970s, the status of human organs became the focus of legislators and the public in the 1980s. The conflict over the commodification of human organs has intensified over the years and is currently an urgent international issue.

Trading Flesh Around the Globe

The Christmas Day, 1983 issue of the *New Jersey Times* carried the following advertisement:

KIDNEY FOR SALE
From 32 yr. old Caucasian
female in excellent health
Write P.O. Box . . .[12]

Several years earlier the *Los Angeles Times* ran the following ad:

Eyes for sale or transplant.
$50,000 each—help someone you care for see and in return
you'll be helping others.
Only sincere parties apply please . . .[13]

These were not isolated cases. In the early 1980s, a man in Georgia offered to sell a kidney for $25,000 in order to buy a fast-food restaurant. Others in the state had offered kidneys for as low as $5,000. The going rate for kidneys in California was considerably higher—some would-be kidney vendors were asking up to $160,000. One of the most publicized attempts to market organs was initiated in 1983 by Dr. H. Barry Jacobs. Jacobs attempted to found an International Kidney Exchange, which would purchase organs from around the world and sell them at a profit to those needy patients who could afford the price. The national uproar about Jacobs's scheme was, in part, responsible for the U.S. Congress passing the National Organ Transplant Act (NOTA), a historic prohibition on organ sales for transplantation.

NOTA, passed in 1984, prohibits the sale, in interstate commerce, of organs or organ subparts. Citing the unique potential of organ selling for exploitation of the economically disenfranchised (who might sell an organ to pay off a mortgage or to feed their children), and the problem of obtaining quality organs from paid donors, NOTA calls for penalties of up to five years' imprisonment and up to $50,000 in fines for anyone caught in "the buying and selling of human organs."[14] As noted by then Senator Albert Gore, who played a key role in shepherding NOTA through to passage, "People should not be regarded as things to be bought and sold like parts of an automobile."[15]

NOTA did not fully prohibit the sale of organs. The bill does not forbid the sale of organs for research, it only applies to organs used for transplantation. Nor did the bill prohibit the sale of all human materials. The bill exempts "replenishable tissues such as blood or sperm" from its provisions.[16] Whatever its shortcomings, NOTA does represent a historic check on the body parts market, and as we will see, it has become a target for those seeking to reinstitute a market for organs.

In 1989, another attempt to establish an international organ business was initiated, this time by a West German businessman who tried to launch a European market for kidney sales from a British base. He described his potential source of supply as "businessmen who have a certain standard of living which they wish to keep or improve and who are willing to sell a kidney to achieve it."[17] In that same year, the British government outlawed the sale of organs, prompted in part by the publicized story of a Turk who claimed that he had been lured to Britain with a job offer, only to find himself hijacked to a hospital where one of his kidneys was removed for transplantation into a wealthy client.[18] Legislators in Germany are pressing for a similar prohibition, spurred by outrage over a Soviet medical institute's offer to provide German patients with Russian kidneys for a fee of $68,570—payable in deutsche marks.[19]

Despite controversy and prohibitions, tens of thousands of organs are being bought and sold around the world. Organs are for sale in India, Africa, Latin America, and Eastern Europe. Donors sell the irreplaceable to buy food and shelter, to pay off debts, even to get a college education. Currently, kidneys in Egypt can sell for $10,000 to $15,000, or the equivalent in televisions and other electronic goods.[20] In India the going rate for a kidney from a live donor is $1,500; for a cornea, $4,000; for a patch of skin, $50. Renal patients in India and Pakistan who cannot find a relative to donate a kidney are permitted to buy newspaper advertisements offering living donors up to $4,300 for the organ.[21]

In India, a recent survey found that a majority of paid "donors" are poor laborers for whom the price paid for an organ could be more than they could save in a lifetime. One donor who set up a modest tea shop with the money paid for his kidney commented, "I am even prepared to sell one of my eyes or even a hand for a price." A mother of two who sold one of her kidneys after her husband lost his job stated, "There was only one thing that I could sell and still keep my self-respect: my kidney." India has found the organ bazaar to be highly profitable. Bombay is reported to be "crowded with wealthy Arabs who buy the kidneys at any price and have them transplanted in $200-a-day clinics or hospitals." Madras is the favorite treatment spot for Singaporean and Thai visitors needing organs. Although India's legislators have failed to act to prohibit the trafficking in organs, many in the Indian medical community are distraught over the organ market. Dr. V. N. Colabawala, an eminent neurologist at Bombay's Jaslok Hospital, states, "We have opened the floodgates to a trade that sacrifices personal morality to expediency."[22]

International reports of organ-selling scandals have become routine. Recently, the *British Medical Journal* described a gruesome scheme to remove and sell blood, corneas, and other organs of patients at a state-run mental hospital in Argentina. According to the *Journal*, the Montes de Oca Mental Health Institute, near Buenos Aires, was removing and selling the organs of its patients. Allegedly, after killing patients for their organs, the Institute would report to relatives that the patient had escaped or died. From 1976 to 1991, the Institute reported over 1,400 "escapes" and nearly as many deaths. However, after complaints from relatives, the Institute was investigated and authorities recovered the remains of the several purported escapees—including the corpse of a sixteen-year-old boy whose eyes were missing. The head of the hospital and eleven others were subsequently arrested.[23]

In 1991, the World Health Organization (WHO) reported that organ selling in the Third World had reached "alarming proportions." The organization has strongly urged member nations to ban the sale of organs. "It is a burning issue for us," said one WHO official, "and we are trying to decide how to deal with it."[24] European health ministers are equally alarmed about the establishment of a market in organs. In 1987, they issued a report stating:

> Organ donation is undoubtedly a profoundly humane gesture, but its legislation and use without major restrictions involve one of the greatest risks man has ever run: that of giving a value to his body, a price to his life. Very many countries, be they poor or very rich, are . . . confronted with the increasing development of an organ market.[25]

Despite the worldwide call to end the market in organs, many in the United States are actively calling for a renewal of organ sales.

The market proponents have suggested setting up a variety of profit incentives to increase the supply of organs. A recent editorial in *USA Today* noted the dire need for more organs, and advocated the payment of a financial reward to families of deceased donors. The newspaper hoped that such a "death benefit might provide incentives that altruism alone could not."[26]

Legal scholar and market advocate Lloyd R. Cohen agrees. He recommends a system in which "organ" contracts are offered to the general public. The contracts would provide that if, at the time of the organ seller's death, organs were successfully taken from the vendor's body and transplanted, a substantial amount of money (perhaps $5,000 per major organ) would be paid to a person designated by the deceased donor. The hospital would be required by law to preserve the vendor's cadaver in a manner suitable for organ harvesting. Cohen is enthusiastic about the market approach:

> Markets are most effective at transferring goods from low-valued uses to high-valued ones. And I can think of no good that fits that category better than a cadaveric organ. The difference in the value of a kidney to the dead versus a kidney to the ill means that there is an enormous price range over which a mutually satisfactory transfer can take place.[27]

Organ sale advocates also argue that a market in body parts will provide individuals with expanded controls and rights over their own bodies. Author and attorney Lori Andrews of the American Bar Foundation supports regulated sale of body parts:

> [A] concern of mine is that in this era when the Supreme Court is chipping away at the right to privacy and when many of us must entrust parts of our body to our physicians . . . we don't really have a way for a person to maintain control over what happens to his or her body parts after they've been removed. I think property is a place to start.[28]

Others are more direct. Physician Jack Kevorkian, notorious for his part in pioneering "assisted suicides," states, "Body parts are property. The person owns them and has the absolute right over what will be done with them in every situation."[29] Some organ market enthusiasts have even suggested that the selling of organs is a legitimate way for the poor to better their condition and that of their families. Underlying each of these calls for a market in organs is the modern faith that the law of supply and demand will somehow create an equitable and just solution to the organ-selling dilemma. As British ethicist David Lamb notes, worldwide attempts to market organs "usually reflect a belief in the alleged efficiency of market transactions."[30]

Despite the siren call of the market, long-standing arguments against selling organs are compelling. Undoubtedly, a variety of profit incentive approaches to obtaining organs would increase supply. However, critics of the market approach note that the sale of organs represents a significant devaluation of society's concept of the body. Dr. Renee C. Fox of the University of Pennsylvania is alarmed by what she sees as the striking "economization" of our language about organ donations and the body. "Discourse about the 'gift of life' and the altruism it entails has given way to discussion about the transplantation 'industry,' the organ 'supply,' 'demand' and 'shortfall,' and about the scarcity of 'human body parts'—often referred to simply as 'HBPs.'" Fox feels that resisting the marketing of HBPs is of great importance to both society and the practice of medicine:

> It is neither accidental nor gratuitous that from its inception, human organ transplantation has been based on the belief that the human body and the extraordinary generosity inherent in donating organs are altogether too precious to be commodified. . . . Because transplantation is institutionalized around the concept that donation is a gift of life, even though the process involves invading and using the body in ways that violate important taboos, donation has attained high moral status and transcendent meaning. Its very legitimacy and what it stands for derive from its association with the values of altruism, solidarity and community. These are not only quintessences of the vocation of medicine, but also of what it means to belong to human society and to contribute to it in a self-surpassing way.[31]

Additionally, as indicated by the growing number of reports from the Third World, appeals to the "market" are often simply euphemisms for the poor selling organs to the rich. Rita L. Marker, director of the

International Anti-Euthanasia Task Force, writes, "Call it what we may, payment for organs is a bounty placed on the bodies of those whose families are least able to withstand financial pressure. . . . It will be the poor, the desperate and the disadvantaged whose loved ones will be worth more dead than alive."[32] Others have noted that if we accept the poor selling organs to escape poverty, why not increase organ supplies further by offering other powerless or oppressed people a "way out" by selling or bartering their irreplaceable body parts? One proposal includes offering long-term prison inmates reduced sentences for donating organs. Most would agree that efficiency in gaining organ supply cannot be the basis of public policy if we are to remain civilized. The ends do not justify the means.

Freedom of choice is also not an "absolute" that can justify the sale of organs. It is an accepted legal and moral imperative that societies have a duty to maintain a community sense of values that prevents egregious exploitation of individuals. Community actions to this end are acceptable, even if such community measures limit individual choice. There are certain forms of exploitation that even a market-driven system does not allow. We do not allow people to sell themselves into slavery. We forbid prostitution. We do not allow people the "right" to sell their labor below a minimum wage. We limit the choice of employers by outlawing job discrimination, be it based on race, creed, gender, or disability. Critics of organ sales persuasively argue that a market on organs fueled by the desperation of the economically disenfranchised easily fits this category of over-exploitative practices that must be prohibited.

Finally, just as arguments for the sale of organs rest on a faith in the benevolence of the market, arguments for the prohibition of such sale are ultimately based on a traditional sense of the respect owed to the human person. It is felt that the integrity, indeed the very definitions of our bodies and ourselves, is changed by a free enterprise in body parts. As ethicist William May notes:

> If I buy a Nobel Prize, I corrupt the meaning of the Nobel Prize. If I buy an exemption from the draft, which was permitted in the Civil War, I corrupt the meaning of citizenship. If I buy and sell children, I corrupt the meaning of parenthood. And if I sell myself, I corrupt the meaning of what it is to be human.[33]

Redefining Death

As the international controversy surrounding the marketing of organs heats up, the demand for these scarce and valuable medical commodities has contributed to a historic and alarming shift in the legal meaning of death. From the earliest times, death in the Western tradition was thought to occur when the soul left the body. In early Jewish teachings, the ability to breathe is given as the *esse* of life. The words of Genesis are clear: "The Lord God formed man out of the dust of the ground, and breathed into his nostrils the breath of life; and man became a living soul." Cessation of breathing was the major criterion of death—a sure sign of the departure of the soul—for most of our history. It was only after Harvey's discovery of circulation in the seventeenth century, and the ability to monitor the heart in the eighteenth century, that the definition of death changed from a primarily spiritual concept to a biological one.

By the twentieth century, the heartbeat and spontaneous breathing criteria of death had become a constant. The classic death scenes now involved a doctor, stethoscope intently pressed to the breast of the dying, frantically attempting to detect a heartbeat; or where a doctor was unavailable, a relative placing a mirror in front of the nostrils of one feared dead in the hope of seeing some faint sign of breath.

By the late 1960s, however, rapid advances in the technology of treating the dying changed the picture of death dramatically. Patients who seemed to have lost all brain function could now be kept alive by life-support systems, which artificially maintained respiration and circulation. Families often were confronted with a loved one who was a late-twentieth-century corpse, part of a tragic new class of the dead sometimes called "beating heart cadavers." These were people whose brains had ceased to function, but whose bodies were kept breathing and pulsating by machines. The dreaded personal and legal decision was whether and when to "pull the plug" and let such a patient die. While these patients created heart-rending dilemmas for families, they were a potential boon for those seeking organs. The artificial maintenance of these lives allowed for an extended time to retrieve organs. Life-support breathing and circulation ensured that organs would still be in good condition when removed.

However, to "harvest" organs from these patients required a new definition of death. A definition based on the cessation of breathing and of heartbeat would not do. To gain viable organs, physicians needed to remove organs while circulation and respiration were still ongoing. Faced with this dilemma, a 1968 Harvard Medical School committee made a historic proposal. They recommended criteria to determine death based on brain activity. The new brain-oriented criterion for determining death is commonly referred to as the *whole brain criterion*. Brain death means that the brain and brain stem have irreversibly and totally ceased to function, and that this will inevitably lead to the permanent cessation of respiration and heart function.[34]

This new criterion for death provided three major advantages. First, in the hands of competent physicians, the diagnosis of irreversible loss of all brain function is said to be clinically practical and reliable. Second, the medical situation for such patients is viewed as hopeless; reportedly, these patients never regain consciousness, and suffer cardiovascular collapse and cardiac arrest within hours, days, or in rare cases, weeks. Finally, as noted, these patients are an excellent source of organs. Under the brain death definition, surgeons could remove organs from patients who were "dead," but whose hearts were still beating and whose lungs, albeit artificially, were still breathing.

By 1981, the American Medical Association, the American Bar Association, and a White House commission had all joined to endorse the whole brain death concept. Within a short time, most states had passed legislation officially ratifying brain death. Now the heartrending dilemma of intensive care units had changed. As noted by Dr. Willard Gaylin in his revolutionary 1974 article, "Harvesting the Dead," "If it [brain death] grants the right to pull the plug, it also implicitly grants the privilege not to pull the plug."[35] The "living cadavers" could be kept alive long enough for organ harvesting. Kin of those who were terminal and on life-support systems could be assured, at some point, that their loved one was indeed dead, that the patient's brain had ceased to function. Then the issue became, not turning the machines off, but rather the painful decision to leave them on for the period of time necessary for efficient organ retrieval to take place.

Today, brain-dead patients can only be kept viable as organ donors for a short time. But as technology improves, many think these "dead" could be kept artificially breathing and functioning for months, even years. These newly named *neomorts* could then be used as whole-body "storage systems" for scarce organs and blood supply, and as research "tools" to test drugs and experimental medical procedures. Dr. Harold Shane, who has written favorably about the neomort concept, admits that "It's a highly controversial idea." He adds, "But technology and wisdom are in a foot race right now, and we need to acknowledge that and move ahead with discussions."[36] Some polls have indicated that Americans are split down the middle on the appropriateness of the neomort concept.[37] While scenarios of the future are debated, changing the definition of death has already markedly increased organ supply. In fact, brain-dead patients are the only source for hearts and livers and are a major source of kidneys. Now, approximately 98 percent of all actual organ donors originate from intensive care units.[38]

The brain death concept, however, has not been an unquestioned success. Even after a decade of use, the concept continues to perplex the public. Many think that brain death includes virtually anyone who is in a coma, regardless of how much brain activity is continuing in the individual. More important, the whole brain death criteria has led to massive confusion in the medical profession, even among those most likely to be making decisions on who is alive and who is dead. A recent survey of 195 physicians and nurses likely to become involved in organ procurement for transplantation came to some astounding and alarming conclusions. Only 35 percent of those making brain death decisions were able to correctly identify the legal and medical criteria for determining brain death. Additionally, well over half of those surveyed "did not use a coherent concept of death consistently." Almost one in five of those surveyed had mistaken concepts of brain death that would have resulted in live patients being identified as legally dead.[39]

Beyond the shocking lack of consistency in medical application of brain death, there are objections to any expansion of the definition of death that is based on the need for a greater supply for organs. Many in the religious community—including Orthodox Jews and a number of Native Americans—believe patients are alive until allowed to die "naturally." Additionally, it is argued by many in the medical community that if it is permissible to change the definition of death because you need organs, then you can move death back earlier and earlier. And their concerns about the further expansion of the definition of death appear to be prophetic.

The Living Dead

Over the last few years, there has been a growing effort to further widen the legal definition of brain death. Current proposals go considerably beyond whole brain death to include among the legally dead those who have lost their "higher" brain functions. This is often referred to as cerebral death, or (somewhat incorrectly) as neocortical death. The idea behind this new definition of death is that those who have suffered a "permanent loss of personhood" or personal identity have actually lost life. Many consider this expansion "an inevitable next step."

If this new concept becomes public policy, those who have lost "higher brain" function but still have "lower brain" function, which often allows them to breathe on their own, would now be included among the legally dead. These patients, first widely publicized through the Karen Ann Quinlan case, are said to be in a "permanent vegetative state (PVS)." This creeping definition of brain death would not only include PVS patients but would also encompass certain newborns. Between one thousand and three thousand anencephalic children—babies born with most of their brains missing—are born in the United States each year.[40] The organs from these children deteriorate quickly after their death. Therefore, certain states are considering legislation declaring these babies "dead" before they die, so that their organs can be efficiently harvested. On October 16, 1987, Dr. Leonard Bailey of Loma Linda University Medical Center in Southern California jump-started the controversy by transplanting a heart from a Canadian anencephalic infant, Gabriel, into a California infant, Paul. This was the first such transplant operation in the United States.

Then in March 1992, the media reported, on a daily basis, the death watch of Baby Theresa, an anencephalic infant whose brain had failed to develop beyond the stem, which controls activities such as breathing and heartbeat. Her parents, who knew about her condition during pregnancy, opted not to abort. They decided instead to give Theresa birth in order to donate her organs to other needy infants. They wanted something good to come from their tragedy. There was a problem, however. If doctors had to wait until Theresa died, there was a good chance her organs would no longer be fit for transplant into other infants. After doctors refused to agree to vivisect the infant at birth, the parents went to court asking for a legal declaration that Theresa was "dead" when born so that her organs could be removed, even though she was still capable of brain stem activity. The Florida Supreme Court in Tallahassee preliminarily rejected the parents' petition on a technicality. Ten days after her birth, Baby Theresa died.[41]

Many were outraged by the court's failure to quickly redefine brain death and allow Theresa's organs to be used. They noted that the child had no higher brain functions and thought little of the argument, used by many, that the baby was owed a life no matter how short or incomplete. Dr. Robert J. Levine, Yale professor of medical ethics, stated that an anencephalic baby "has more in common with a fish than a person."[42] Levine's view has some support from legal commentators. For several years, a number in the legal profession have supported "recognizing that children born with anencephaly are not alive."[43] Others, however, were vocal in their support of the concept that the anencephalic child was not "dead." Columnist Charles Krauthammer wrote:

> It is tragic that one cannot use the organs of a hopelessly doomed anencephalic to save the lives of other infants. But to kill one innocent for the sake of another is simple barbarism. Even just shortening a doomed life in order to lengthen another is a fateful start on that road. And it will not stop there. The anencephalic is the frontier case, and the frontier is always moving. Next comes the irreversibly comatose adult . . . then come the Alzheimer's patients. Why not bring some good out of their tragedy too?[44]

Once again, as in whole brain death, the major rationale for expanding those defined as dead to include those who have lost "higher" brain function, or "personhood," is the need for more organs. Over 2 million people die each year in the United States, but even under current whole brain death criteria, only about 25,000 of them are suitable sources of organs.[45] The principal requirements for this select group of dead suitable for organ harvesting are that the donors died in good health and died suddenly, as in a traffic accident or from a stroke. This limited legitimate pool of organs means that transplantation can never—even under optimal retrieval circumstances—be a serious solution to heart, kidney, liver, or lung disease. At best it could add a few years of life to only a small percentage of those dying of these diseases.

Expanding the definition of death to include those who have lost "higher" brain function could add tens of thousands of "dead" to those who are suitable organ donors. Equally important, many of those who have only lost higher brain function could be highly dependable long-term sources for organs. Because they breathe on their own and do not require artificial respiration as current "brain dead" patients do, they can be used as neomorts far more easily. As explained by Dr. Robert Smith, an advocate of expanding the definition of death:

> A neocortical death standard could significantly increase availability and access to transplants because patients (including anencephalics) declared dead under a neocortical definition could be biologically maintained for years as opposed to a few hours or days, as in the case of whole brain death. Under the present Uniform Anatomical Gift Act, this raises the possibility that neocortically dead bodies or parts could be donated and maintained for long-term research, as organ banks, or for other purposes such as drug testing or manufacturing biological compounds.[46]

Many object to a future vision of thousands of breathing neomorts in hospitals around the country being used as living "banks" of organs for donation and experimentation. These critics of the "higher" brain death concept point out that there is no dependable scientific criteria for what is termed "higher brain" function and "lower brain" function. They also argue that, though rare, there are many reports of recovery from permanent vegetative states. These reports dramatically indicate the lack of certainty in diagnosing irreversible absence of awareness.[47] While this uncertainty remains, organ removal from permanent vegetative state patients, and similar patients who have lost higher brain function, could mean vivisecting the living. For many, this prospect throws into doubt the whole transplantation enterprise. As stated by the Conference of European Health Ministers in 1987, "It would be preferable by far for man's future survival to have to abandon transplantation than to agree to remove vital organs from individuals who are not really dead."[48]

Even assuming surefire diagnosis of irreversible "higher" brain death, treating people in a permanent vegetative state as cadavers creates a web of ethical dilemmas. As ethicist David Lamb maintains:

> The notion of a still breathing corpse is morally repugnant. How, for example, does one dispose of such a being? Should burial or cremation take place whilst respiration continues? Or should someone take responsibility for suffocating the "corpse" first? And what would be the outcome if a distraught family member suffocated a relative who had been vegetative for months? Would it be homicide? Or would it be seen as unacceptable treatment of a corpse?[49]

Additionally, as noted by Krauthammer and others, there is legitimate concern that shifting the definition of death to include "higher brain" function will lead to yet a further expansion of death in future years. One prominent transplant expert states:

> I am opposed to "higher brain" formulations of death because they are the first step along a slippery slope. If one starts equating the loss of higher functions with death, then, which higher functions? Damage to one hemisphere or both? If to one hemisphere, to the verbalizing dominant one, or to the "attentive" non-dominant one?[50]

Neonatologist Dr. Jacquelyn Bamman has similar concerns about infants. She is opposed to using anencephalic children, prior to their natural deaths, as donors for transplants and is concerned about things to come, including using the expected life span of babies as a criterion for declaring death and harvesting organs. If all infants with short life spans or whose brains are severely abnormal are potential "donors," Dr. Bamman observes, the field is opened to infants with hydrocephaly, grade IV intracranial hemorrhage, Trisomy 13 and 18, and a host of other handicaps.[51]

As in the controversy over the sale of organs, our view of the human body is key to decisions on how to define death. Ultimately, questions of what is life, what is death, who is a proper donor of organs and who is not come down to our understanding of what is human and what constitutes personhood. Despite the views of many in the medical community, definitions of death cannot be separated from moral and ethical judgments. Many see those who are "higher" brain dead as little more than efficient sources of organs and as potential tools in biomedical research. They do not view these humans as alive or meaningful. But if brain activity is not the only criterion for respect due to humans who are still breathing and otherwise functioning, then neither those in the so-called permanent vegetative state nor anencephalic babies are dead. Their personhood, as reflected in their living bodies, is still functioning. These patients cannot be treated as mere receptacles of a valuable variety of body shop commodities, no matter how altruistic the result. If tests reveal that there is no hope for those who have lost certain brain functions, they must still be treated with care and respect until they expire. No physician should be allowed to vivisect them for organs. They cannot be treated solely as means. They are still ends in themselves.

There is, in fact, something absurd and ethically myopic about the attempt to say that a person can die, but that person's body cannot. The new definers of death have created a "ghost in the machine" scenario. Psychiatrist Dr. Stuart Youngner, a supporter of using "persistent vegetatives" for body parts, is clearly a victim of this false dichotomy. He states, "After the loss of personhood what remains is a thing, the demise of a body that has outlived its owner."[52] The radical separation between consciousness and body promulgated by adherents of many brain death theories leads the new definers of death to degrade the body as little more than a bank of spare organs and other biological materials. Philosopher Hans Jonas sums up the objections to the new brain death concept:

> I see lurking behind the proposed definition of death, apart from its obvious pragmatic motivation, a curious revenant of the old soul-body dualism. Its new apparition is the dualism of the brain and the body . . . it holds that the true human person rests in (or is represented by) the brain, of which the rest of the body is a mere subservient tool. . . . [Yet] my identity is the identity of the whole organism, even if the higher functions of personhood are seated in the brain. . . . Therefore, the body of the comatose, so long as—even with the help of art—it still breathes, pulses, and functions otherwise, must still be considered a residual continuance of the subject . . . and as such is entitled to some of the sacrosanctity accorded to such a subject by the laws of God and men. That sacrosanctity decrees that it must not be used as mere means.[53]

Just as transplantation has caused profound confusion about the definition of death, it is also becoming part of the ongoing explosive debate on the definition of life. As we will see in the next chapter, the unborn are now joining the born as sources for organs. Fetal organs are increasingly being used in human transplantation, and as transplants into animals to create more effective research subjects. The controversy surrounding fetal transplants, like the potential fetal organ market itself, may dwarf that of the current organ transplant industry by the next century.

3

The Gods themselves cannot recall their gifts.
Alfred, Lord Tennyson, "Tithonus"

Harvesting the Unborn

THE ANATOMY "SPECIALIST" for the International
Institute for the Advancement of Medicine (IIAM) in Exton, Pennsylvania, has a grim daily round. He travels to local abortion clinics in
his region, scavenging for abortion remains. He is in search of fetal
parts, ranging from whole hearts to brain slivers and other organ
fragments—any fetal tissues that may be valuable to doctors and
researchers. At any one time, the Institute has as many as twelve
"specialists" harvesting fetal organs at abortion clinics in Pennsylvania. They garner organs and tissues from fetuses up to twenty-four
weeks gestation. The Institute's collectors "harvest" approximately
seven hundred specimens per month from 450 second-trimester
abortions. IIAM procures first-trimester aborted fetal parts less frequently. Altogether, the Institute procures over eight thousand fetal
parts each year.[1]

To harvest valuable fetal organs, IIAM has arrangements
with eighteen abortion clinics, mostly those that perform second-
trimester abortions. The Institute pays the abortion clinics a "service
fee" for allowing it to search for and take fetal parts. In turn, IIAM

charges a fee for the fetal parts it distributes to researchers. Buyers of IIAM's tissues pay "handling fees" of between $50 to $150, depending on the fetal tissue specimen. For its customers, IIAM attempts to keep charts on the fetus's medical and family history. However, testing the fetal parts for infectious diseases such as HIV and syphilis costs extra. According to IIAM's administrator, James S. Bardley, Jr., fetal tissue sales bring in close to $1 million annually.[2]

IIAM, established in 1986, is one of a half-dozen identified fetal tissue providers around the country. The actions of these fetal harvesters, and virtually all procurers of fetal tissues, are shrouded in secrecy (IIAM has an unlisted number and will not identify its research clients or those clinics that it uses for supply).[3] These tissue brokers are virtually unregulated; when fetal demand grows, new "seat of the pants" fetal brokers tend to sprout up around the country to supply that demand. Even the federal agencies have difficulty in getting facts about fetal organ harvesting. A study conducted for the National Institutes of Health (NIH) states that "Many informants were reluctant to reveal information about [fetal parts] procurement. They cited the need for confidentiality about research in progress, but they also discussed uncertainties surrounding legal or regulatory requirements and concerns about public opinion."[4]

It is believed that each year tissue brokers like IIAM distribute approximately 15,000 specimens to researchers and doctors.[5] Many university and commercial researchers do not use brokers or tissue banks to obtain fetal parts, but arrange privately with local abortion clinics for these tissues. A random survey by the National Abortion Federation found that 11 of 150 clinics contacted supply fetal remains directly to medical researchers. Hospitals, where approximately 10 percent of all abortions are performed, also are providers of fetal tissues. Larger medical centers, such as those at Yale and the University of California at San Francisco, regularly provide fetal tissue to their own researchers and doctors.[6]

The harvesting of fetal parts is essential to a new research and transplantation industry. The new industry is based on transplanting fetal organs and organ subparts—most often "harvested" from elective abortions—into people, and increasingly into a variety of other animals. Although it is deeply controversial, the use of fetuses for research and transplantation has caused great excitement in the biomedical community. Many scientists and researchers have her-alded fetal transplantation as one of the most promising areas of human biotechnology. Dr. Antonin Scommenga, a prominent scientist in transplant technology, has declared that with fetal tissue use, "[W]e are confronted with a biological revolution which is going to be just as important as the nuclear revolution was for physics."[7]

The excitement about fetal transplantation is not just scientific. There is no doubt that fetal research and transplantation could be big business. Economist Emmanuel Thorne has predicted that the fetal transplant industry could "dwarf" the present organ transplant industry in coming years.[8] Fetal organs and tissues could become one of the most valuable commodities in the human body shop.

The Fetal Revolution

The concept of using fetal tissue for trans-plants and research is not new. Attempts at fetal transplants, and other medical use of fetal tissue, go back over six decades. The year 1928 saw the first unsuccessful attempt to transplant a fetal pancreas into a patient with diabetes. The first U.S. fetal pancreas transplant occurred in 1939 and was also unsuccessful. The first successful use of fetal tissue in medicine was the utilization of fetal kidneys in the development of the polio vaccine in the 1950s. The late 1950s also witnessed sporadic and unsuccessful attempts at a number of fetal organ transplants. By the early 1960s, both the United States and the United Kingdom had set up fetal procurement organizations to sup-ply researchers with fetal tissues.[9]

Whatever the past use of fetal tissue, the last decade of the twentieth century will undoubtedly be seen as a watershed in the "harvesting" and use of human fetal organs for human transplanta-tion and animal research. The fetal "revolution" in America was precipitated in November 1988, when a team of doctors in Denver, Colorado, captured the medical spotlight by announcing that they had performed the nation's first transplant of fetal neural tissue. The doctors, led by Dr. Curt R. Freed, implanted tissue from the brains of fetuses into the brain of a fifty-two-year-old Denver man, Don

Nelson, who was suffering from Parkinson's disease. The fetal transplant involved drilling a quarter-size hole into Mr. Nelson's skull and implanting the fetal brain cells deep into his brain. The aim of the transplant was to have the transplanted fetal tissues begin manufacturing dopamine in Nelson's brain. While the causes of Parkinson's disease are still unknown, the symptoms of the disease are triggered by the destruction of cells in the brain that produce the hormone dopamine. Dopamine helps transmit nerve impulses between the brain cells that control muscle activity, aiding in fluent speech and coordination. Symptoms for Nelson and other victims of the disease include uncontrolled shaking of feet, head, and hands, and difficulty in speech.[10]

In the months after the operation, Dr. Freed reported some improvement in his patient's memory and coordination, though long-term improvement has not yet been fully proved. Since his first operation, Freed has conducted seven other fetal transplants into Parkinson's victims. Yale University has performed eleven similar surgeries in the years following the first Denver transplant. As of June 1992, the nineteen American fetal transplant procedures were part of more than one hundred fetal brain transplants performed in countries around the world.[11] Most notable was the ground-breaking work of Swedish neurologists Olle Lindvall and Anders Bjorkland. The Swedish team, which conducted its implants one year before the U.S. team, was the first to report that fetal cells stayed alive and carried out their function after transplantation into a patient's brain.[12]

Fetal tissue transplants are also being tried as a therapy for diabetes. According to a study by the University of Minnesota's Center for Biomedical Ethics, about six hundred insulin-dependent diabetes patients worldwide—thirty-eight in the United States—were reported to have received fetal pancreas transplants by 1990. These transplants are designed to help patients begin making their own insulin. However, they have proved generally ineffective due to rejection problems. Additionally, about three hundred patients worldwide have received fetal liver transplants; only a few of these transplants have been performed in the United States.

Even ardent advocates are quick to note that the use of fetal transplantation as a cure for Parkinson's, or other diseases, is far from proven. Many early reports about the success of such experiments have been found to be exaggerated and inaccurate. Several recent reports indicate that, as of now, fetal tissue transplants have been largely unsuccessful as a therapy for Parkinson's.[13] "I don't see any study [of fetal brain tissue transplants] that shows anything unequivocally different than what is seen after the adrenal medulla grafts," states William J. Freed, chief of preclinical neuroscience at the National Institute of Mental Health. Freed notes that some reported improvements after fetal grafts could be due to the brain's natural ability to repair itself following an injury or surgery.[14]

Despite controversy over their effectiveness, fetal transplant operations are expected to continue and increase over the next few years. In fact, according to many in the medical community, Parkinson's and diabetes are just two illnesses for which fetal tissue offers a potential cure. Researchers claim that fetal transplantation could eventually cure tens of millions of Americans. Prospective recipients of fetal transplants could include 1 million Parkinson's disease victims, 3 million Americans suffering from Alzheimer's disease, 25,000 with Huntington's disease, and 6 million diabetics.

Future visions for the fetal revolution in biotechnology go far beyond specific cures currently under consideration. Bioengineers envision large-scale use of the unborn as sources for organs and tissues to repair generations of men and women who are getting progressively older. As one scientist stated, "We are on the threshold of changing ourselves. I realize this opens a Pandora's box, a can of worms, or whatever you want to call it, but I foresee growing fetuses someday for spare parts."[15]

Ethicists are already chronicling a growing number of cases where individuals are seeking to arrange their own source of fetal organs by conceiving with the intent to abort. Arthur Caplan, a medical ethics specialist at the University of Minnesota, has cited a case of a woman who proposed being artificially inseminated by her father, an Alzheimer's victim, so genetically matched fetal cells could be used to treat him. In another case reported by Caplan, a woman with severe diabetes wanted to conceive and then abort a child in order to use the pancreas cells from her unborn child for transplants to help her condition.[16] While these and similar requests for fetal tissue use are apparently being rejected by most doctors, no guidelines exist under which the medical community can make these novel ethical decisions. In fact, many experts feel that the current disapproval of "growing fetuses" for medical uses will be short-lived. As noted by economist Thorne, "The potential uses of fetal tissue make the temptation to conceive with intent to abort almost inevitable."[17]

We are already on the threshold. In June 1991, fourteen-month-old "miracle baby" Marissa Eve Ayala was used to provide bone marrow for her eighteen-year-old sister, Anissa, who was stricken with leukemia. The Ayalas, Mary and Abe, gained national attention with the announcement by Mary Ayala that she had conceived Marissa solely for the purpose of providing a compatible bone marrow donor for her daughter Anissa. Beating heavy odds, Marissa was born with bone marrow compatible with her sister's, and the transplant was performed.[18]

Although the transplant posed little hazard to Marissa, the Ayala case brought up several unprecedented ethical issues with significant implications for future "fetal farming." What if prenatal testing had found Marissa not to be a bone marrow match with her sister? Would Mary Ayala have had an abortion and tried again, perhaps several times, to conceive a donor for her daughter? Some leading practitioners and ethicists believe that attempting to create fetus after fetus to find the right tissues is totally acceptable.[19] An investigation subsequent to the Ayala case revealed that dozens of women have reported becoming pregnant solely to gain bone marrow or organs for an already existing child.[20] These cases present a disturbing precedent for those seeking to prevent unrestricted fetal use. They underscore the growing impetus among the medical community to allow for conception of fetuses solely for their use for transplantation.

"Humanized" Mice

Fetal transplantation in the United States involves not only fetal transplants into humans, but also into animals. Cross-species organ transplantation came to the world's attention in 1984, when a surgeon in Loma Linda, California, inserted the heart of a baboon into an ailing infant girl, Baby Fae, in a futile and highly questionable attempt to save her life. In October 1990, immunologist Dr. J. Michael McCune made headlines with a bold if bizarre inversion of the Baby Fae operation. Dr. McCune initiated a group of experiments successfully transplanting fetal organ subparts into laboratory mice. In these experiments, tiny human organ structures, including lungs, pancreases, and intestines, were implanted into mice born without immune systems. The experiments had little to do with helping the mice; the transplants were performed in order to make the mice more effective research tools for testing a variety of human drugs.[21]

To create mice with human immune systems, Dr. McCune takes the thymus, liver, and lymph nodes from human fetuses under twenty-two weeks old. He then implants a piece of each organ under the kidneys of young mice. Within a few days, the mouse's blood vessels invade the miniature organ subparts and the fetal organs begin to grow, eventually engendering cells of the human immune system. Once the immune system is in place, these mice, now called "humanized mice," are infected with the AIDS virus or with leukemia-causing viruses. The animals are then used to screen a variety of antiviral compounds for their effectiveness in fighting these diseases. McCune believes he has thereby created a "human" disease model in nonhuman mammals. The contract testing of HIV antiviral drugs on these mice is already underway on behalf of several pharmaceutical companies. Systemix, the company for which McCune acts as research director, offers two basic programs: one smaller screening program that uses about 25 "humanized" mice (cost, approximately $25,000), and a larger one that involves about 160 of the novel mice (cost, approximately $65,000).[22]

McCune got what he calls his "wild ass" idea for creating a mouse with a human immune system in January 1987 while driving home to San Francisco from Stanford University, where he studied immunology. Among the technical problems with creating a humanized mouse was the realization that the human materials would tend to reject their nonhuman host, imperiling the mice. As McCune contemplated this difficulty during his drive, he suddenly got a flash of inspiration: aborted fetuses. The organs of fetuses generally have not developed immune hostility to foreign bodies, so no rejection would take place. McCune tested his hypothesis and found that it worked. Moreover, as his research continued, McCune discovered that he did not appear to encounter any opposition from antiabortion or animal rights groups. Present plans for Systemix include implanting other fetal organs into mice, including lungs, intestines, pancreases, pituitary glands, skin, brain cells, and placentae.[23]

McCune's fetal transplant work, though primarily funded by the National Institutes of Health, is not solely altruistic. Success could mean big money for Systemix. As of 1991, the enterprise's main investor—to the tune of $10 million dollars—is Eli Jacobs, the owner of the Baltimore Orioles baseball team. As the Systemix enterprise gathers momentum, several other researchers and corporations have also begun transplanting fetal organs into laboratory animals.[24]

Why Fetal Parts?

Fetal tissue has several attributes that make it ideal for transplantation into humans and animals. These attributes allow for experiments like those of Freed's and McCune's and provide some basis for the view that fetal use will be the "nuclear revolution" of biology. As noted, these same advantages make it an ideal commodity for the growing biotechnology body shop. First, although fetal tissue does not come from what the Supreme Court in *Roe v. Wade* held to be constitutionally recognized "persons," it is unquestionably human. This is of great importance in easing the problem of rejection that plagues many transplant procedures. Clearly, for human recipients, human tissue and organs are immunologically more compatible than biological materials from either other animals or artificial models, and are less likely to be rejected. In fact, the ideal solution for any transplant recipient is to obtain biological materials from his or her own body or that of a close relative, a feat that can be achieved by using fetal parts. Moreover, as McCune's mouse experiments show, fetal tissue does not reject being transplanted into nonhuman hosts. Fetal tissue is available for both human and nonhuman transplantation.

Of almost equal importance is the fact that fetal tissue exhibits tremendous developmental potential. The fetus is, after all, a growing organism, and this potential is passed on to fetal body materials. Generally, fetal organ subparts—whether growing in the biotechnologist's lab or transplanted in an animal or human recipient—grow far more easily than mature human biological materials.

Another clear advantage of fetal materials, as contrasted with other human organs, is the abundant available supply. Over 1.6 million fetuses are aborted each year in the United States, over 30 million worldwide. As we will see, this number represents fetuses only potentially available to the biotechnology industry. Actual use of most of these electively aborted fetuses for organ harvesting will require significant changes in current abortion practices, including wholesale alterations in the method, timing, and manner of abortions. Induced abortions also allow scheduling of fetal material "harvesting." Substantial evidence indicates that fetal tissue is only effective for as long as it remains alive—though research into creating "banks" of frozen fetal organs is ongoing. Since elective abortions, by definition, occur at a scheduled time, it is possible for researchers to obtain tissue in the shortest interval possible, which increases the likelihood that the tissue will be alive at the time of transplantation.

Given the advantages of fetal tissue for engineering human beings and making research animals more efficient, plus the clear market potential of fetal transplants, many are puzzled by attempts to slow or stop fetal research and transplantation. Judy Rossner, director of the United Parkinson's Foundation, notes that if fetal tissue works, it should be used. She has little patience for critics. "We live in a very Victorian society," Rossner explains. "The religious people can go bury their heads in the sand for all we care."[25] *Medical Technology Stock Newsletter* editor Jim McCamant sees morality getting in the way of a profitable and effective technology. "Using fetal tissue is not a moral question for me," McCamant writes. "I believe this technology can work." Congressman Henry Waxman, a major booster of fetal tissue use, states that people opposed to fetal transplants are guilty of "disinformation," "bad medicine," and "dangerous science policy."[26]

Unfortunately, for those who wish to impel the world of fetal research and transplantation into the hospital and marketplace, each advantage of fetal tissue raises unique and complex questions—ethical dilemmas that will not disappear. In fact, the concerns over fetal transplantation starkly contrast utilitarianism with reverence for life, market values against traditional sensibilities, and bring into question the very definitions of life and death.

Objections to the use of fetal tissue derive from several considerations. Perhaps foremost is the question of abortion itself. Clearly, many of those opposed to elective abortions will also oppose all use of fetuses obtained by such abortions. These concerns are certainly understandable for those opposing abortion. Once a steady supply of fetal parts becomes a staple of a growing and lucrative transplantation industry, it appears likely that medical institutions would become closely allied with the cause of continuing elective abortions, if only on a self-interest basis. Ethicist Steve G. Post of Case Western Reserve University writes,

> With the advent of widespread fetal tissue transplant, elective abortion would no longer be a political issue that biomedical researchers, as scientists, could ignore. Rather livelihoods and institutional grants would demand that elective abortions be continued. I wince at the idea of biomedical science entering the abortion fray.[27]

However, though the media has focused almost exclusively on the abortion aspect of the fetal tissue controversy, other profound public policy issues remain. Of primary concern is the propriety of sale of fetal parts by women, clinics, and hospitals. The selling of fetuses places women as manufacturers of a new commodity in the human body shop, with clinics and hospitals as traders and utilizers of the product. The concern is hardly abstract.

The Fetus Market

In 1989, bronze star Vietnam veteran Craig McMullen found himself in a new kind of mine field. The balding, bespectacled McMullen was the president and chief executive of Hana Biologics in Alameda, California. Hana, formed in 1979, originally manufactured various formulas for the biotechnology industry. But business was slow. In 1984, McMullen joined the firm, and within the next year Hana had refocused its efforts toward fetal cell research. As one of only a handful of biotech companies devoting itself to fetal transplantation, Hana soon attracted hordes of securities analysts to its presentations, where the company boldly outlined its plans to cure diabetes, Parkinson's disease, and hemophilia by using fetal tissue. In the case of diabetes, the company was harvesting insulin-producing pancreas cells from aborted fetuses, hoping to grow these cells and then implant them into diabetes patients, relieving them of the need for a lifetime of insulin injections. The company was getting more than its share of press attention.[28]

The pot of gold awaiting Hana if it could perfect fetal transplantation technology and overcome the controversy surrounding fetal research was considerable. The market for fetal products to cure diabetes was estimated at $3 billion, Parkinson's disease at $3.5 billion. Alzheimer's, spinal cord injuries, and various blood diseases also offered potential markets of major proportions.[29] However, Hana faced several obstacles on its way to marketing fetal technology. As early as 1986, the company had been concerned about the implications of the National Organ Transplant Act (NOTA) on its business. As described earlier, NOTA forbids the sale of human organs for transplantation. However, it makes no mention of fetal organs or organ subparts. Lawyers for the company were confused about how NOTA might affect the ability of Hana to sell its fetal products.

Even worse for the company, public interest groups were becoming aware of Hana's attempts to sell fetal parts for transplantation. Right-to-life demonstrations in front of Hana's offices were not unusual. In September 1987, a biotechnology watchdog group, the Foundation on Economic Trends, petitioned the Secretary of Health and Human Services (HHS), Dr. Otis R. Bowen, to include fetal organs and organ subparts as part of the protected class of organs under NOTA. The petition, specifically describing the activities of Hana, formally requested that the HHS "immediately designate the subparts of fetal organs and all non-regenerative fetal tissue as prohibited organs under the Organ Transplant Act."[30]

The Foundation's petition caught HHS in an awkward ideological dilemma. The Reagan administration was torn between the ideology of the free market, upon which it had launched its deregulation revolution and which would allow the open sale of fetuses, and the traditional ethical understanding, undoubtedly shared by many in the Administration, that nonreplaceable human materials and especially fetuses should not be part of the commercial market.

Predictably, HHS hesitated. After several months, however, HHS Secretary Bowen, later to reveal himself a strong proponent of fetal tissue use, decided to act. The marketers apparently prevailed. The HHS prepared a response to the petition that turned down the Foundation's request and stated that fetal subparts were not covered by NOTA—they could be sold. However, this reply never was officially released, due to concern that it would lead to difficult and damaging litigation against the HHS by the Foundation. Frustrated by HHS indecision and delay, Foundation attorneys went to Congress to get fetal organs and organ subparts protected from sale in the open market. In November 1988, thanks to the careful shepherding of Senator Gordon Humphrey and his staff, an amendment to NOTA protecting fetal organ and organ subparts from sale was passed by Congress and signed into law.[31]

The publicity surrounding the passage of the NOTA fetal amendment, and the continuing controversy over fetal tissue use, pushed Hana's stock down almost 85 percent. But Hana's and McMullen's ultimate demise had more to do with skewed science than with the deeper issue of the ethics of selling fetal parts. In late 1989, Hana scientists discovered that much of their research had been studded with errors. The fetal pancreatic cells it had been growing, and publicly touting as a cure for diabetes, did not work. The cells were supposed to produce insulin—they didn't. This shocking discovery led to McMullen's resignation and to a layoff of 25 percent of Hana's work force. Hana subsequently shifted its research away from fetal transplantation.[32]

The saga of Hana and the success of legislation to prevent open sale of fetal parts do not end the controversy over the marketing of fetal tissue. First and foremost, as with other human organs, NOTA only forbids sale of fetal parts for transplant purposes. It does not cover sales of these organ and organ subparts for nontransplant research. Additionally, there is considerable legal confusion on whether NOTA includes protection against the sale of fetuses for transplantation into nonhuman subjects, such as the engineered mice of Dr. McCune. Moreover, as described above, how companies or researchers actually obtain their fetal parts from brokers like IIAM, or abortion clinics, is often obscure. Who is making money on fetal parts, and how, is unclear. Many in the fetal tissue procurement industry are undoubtedly attempting to circumvent legislative prohibitions on fetal parts sales by making indirect profits—including so-called service fees, handling fees, and procurement fees—on the unborn. Finally, as described earlier, there is considerable pressure by a growing number of market advocates to repeal NOTA itself and begin an open market, or controlled sale, of all human organs, including fetal parts.

Feminist scholar Dr. Janice Raymond supports abortion but opposes government support for fetal organ transplants. She is concerned about yet another fetal marketing scenario: international sale of fetal parts, a market that could occur regardless of U.S. law. Raymond has calculated that under 10 percent of the more than 1.5 million abortions performed each year in the United States render usable fetal parts, that is, fetal parts that have not been destroyed by the abortion process and are at the right stage of development. She notes that this supply would not be enough to treat all Parkinson's victims or diabetes victims, should fetal transplants be shown effective. She fears this could lead poor women in other countries to sell their fetuses. "I'm raising the specter of international trafficking in fetal parts," Raymond says. "I think that it is very realistic."[33]

Dead or Alive

An important and grisly unresolved issue surrounding fetal transplants is the almost inevitable vivisection of still living fetuses in the process of fetal organ "harvesting." In the early 1970s, there were several documented cases, both in the United States and abroad, of experiments deliberately using live fetuses for research.[34] One macabre 1973 study conducted by a U.S. and Finnish team involved the decapitation of twelve fetuses (up to twenty weeks gestation) aborted alive through hysterotomy. The heads were kept alive by perfusion through the internal carotid artery in order to study cerebral oxidation of glucose. The experiment was partially funded by the National Institutes of Health.[35] In response to public concern about such experiments, Congress passed a moratorium on research involving living fetuses in 1974. Additionally, a number of state legislatures mandated that no fetal research take place on any but dead fetuses.

However, researchers are still harvesting tissues from live fetuses. There are verified reports of modifications of abortion techniques that remove the fetus from the uterus alive and intact in order to make it easier to obtain certain types of fetal tissue. There are also reports of Swedish researchers harvesting tissue from a live fetus in utero prior to it being aborted.[36] James S. Bardley, Jr., director of IIAM, whose practices were described above, openly admits that IIAM looks for doctors who use certain suction methods to obtain fetal parts intact. To harvest fetal parts from second-trimester abortions, in which IIAM specializes, he advertises for doctors who use a technique called dilation and evacuation (D&E), in which the fetus is essentially pulled out of the anesthetized woman. The problem with other methods of abortion is that the fetus is killed prior to inducing labor. Because the fetus is often alive during D&E abortions, Bardley admits that "some doctors are squeamish about D&Es." "However, we need tissue that is fairly fresh," notes Bardley. "We have to process the tissue within minutes of the time of death."[37]

Not all harvesting of organs from live fetuses is intentional. This is due in part to the difficulty of defining death in a fetus. While a variety of medical groups, commissions, and states have tried to define fetal death, these guidelines cannot be confused with dependable diagnostic tests. Brain death is often extremely difficult to ascertain in fetuses and is actually not always possible even in infants up to one week old. One commentator has concluded that "the only certainty about infant brain death diagnosis seems to be that there is hardly any certainty, at least given the current state of art."[38] Neurobiologist Dr. Keith A. Crutcher states that "[t]he application of current criteria for determining fetal death . . . is insufficient for insuring that living (non-viable or viable) fetuses are not subjected to vivisection."[39] Moreover, most physicians involved in fetal organ procurement do not make any serious attempt to ascertain whether aborted fetuses are still alive. Standard confirmation of death by two physicians monitoring the heartbeat or other vital signs of fetuses is not routinely conducted. Moreover, the researchers and doctors who obtain fetal tissue for transplantation from clinics or brokers do not monitor whether the fetal parts they acquired were removed from living or dead fetuses. And, despite numerous requests for investigations, the National Institutes of Health and other federal oversight agencies have virtually ignored the vivisection problem.[40]

In general, researchers involved in fetal tissue use are not eager to deal with the sensitive and difficult issue of fetal vivisection. However, over the last five years, several adherents of fetal tissue use have openly begun advocating that it is permissible to procure fetal tissue from fetuses still alive. Some have recommended that an aborted fetus be classified as "dead" regardless of continuing heartbeat or lung activity after abortion.[41] Others suggest that the fetus be given a new category as an organism "not yet living," which would allow its use as an organ source while respiration was still ongoing.[42] An argument often used by some fetal tissue use adherents is that most abortion procedures result in "nonviable" fetuses—the fetus even if still living after being aborted has no chance of survival. This is almost certainly the case. Fetuses still alive after the abortion process are not going to live for any appreciable time. However, nonviability is not equivalent to death. As noted by Dr. Crutcher, "Depending on the time span used as the reference, we are all non-viable."[43]

Many suggest that the "higher" brain criteria for death described in the prior chapter, not yet accepted for born human beings, be used for fetuses. They posit that even if the nonviable fetus that has been aborted from the uterus exhibits heartbeat and respiration, it has irreversibly lost the potential for "consciousness" and is therefore dead and available for vivisection. Others suggest similarly that the fetus cannot be counted among the living until its brain "begins to function as the control center for the integrated functioning of the organism," a process that does not occur until its twenty-second week; until this time the fetus is considered a "nonhuman."[44]

Even if these attempts to redefine life fail, there will continue to be pressure to vivisect fetuses for spare parts based on efficiency and profitability. As demonstrated by the statement of IIAM director Bardley, as long as there is a significant medical and financial incentive for those procuring fetal organs for transplantation to secure the tissue within the shortest period of time after abortion, there is also an increased likelihood that the aborted fetus will still be alive when dissected. Given the past and current history of live fetal use, proponents of using fetal parts in research and transplantation face the burden of establishing new regulations that would provide appropriate safeguards so that organs would not be harvested from fetuses which are still alive and still capable of feeling pain.

Complicity or Consent

PERMISSION FOR EXAMINATION OR TRANSPLANTATION OF THE TISSUE OBTAINED AT THE TIME OF ABORTION

Scientific studies from tissues derived from abortion are important to the understanding, treatment, and prevention of many birth defects and other diseases. Additionally, aborted material can be used for the purposes of transplantation, therapy, education, and research.

I hereby grant permission to _____ Hospital and/or clinic and each of their authorized agents to distribute and dispense this fetal material for biomedical research and hereby also donate all relevant aborted material as an anatomical gift for the purposes aforementioned. I also agree that a blood sample may be taken before abortion.

_____   _____

PATIENT'S NAME   DATE

_____

WITNESS

This sample fetal organ donation form is typical of many such forms that are increasingly being presented to thousands of American women each year.[45] But can a woman ethically consent to the use of the aborted fetus for science or transplantation? Legal experts are confused about how to apply traditional concepts of informed consent when the death of the donated fetus is willed by the pregnant donor. Legal consent for use of the organs of a cadaver almost always involves an individual's or next of kin's permission to use a body *after* an unavoidable death. It is arguable that such consent is not ethical when the individual giving the consent is also deciding that the death will happen.

Moreover, regardless of one's position on abortion, there exists the potential of fetal transplants acting as an incentive for elective abortions and being a coercive element in a woman's decision on the method and manner of abortion. Though often denied by fetal tissue use proponents, it seems well within the realm of possibility that for women agonizing over the difficult decision of abortion, publicized accounts of the potential medical miracles that might be accomplished with fetal tissue could act as an important factor in the abortion decision. This would be especially true if abortion clinics openly advocated fetal donation in their counseling and in their consent forms.

The altruistic overtones of fetal tissue use could also provide a rationale for researchers to encourage many women to alter the method and manner of their abortions in order to provide "optimum specimens" for research and transplantation. While various national and international ethical guidelines discourage this practice, they have not stopped it. There have been numerous documented instances in which alterations in abortion technique have been used to increase the amount and quality of fetal tissue. These include but are not limited to: increased use of ultrasound to better ascertain fetal placement; reducing the pressure of the suction used to pull the fetus from the womb in order to get fetuses intact; using larger cannula to get less "disrupted" tissue; and using forceps to remove fetal tissue before performing vacuum abortion.[46]

Many of these techniques are more intrusive and hazardous to the women, but better preserve the fetus. For example, modifications in abortion techniques for the purpose of obtaining fetal tissue may lengthen the abortion procedure from less than ten minutes to approximately thirty minutes. These techniques could cause cramping, infections, cervical damage, and other physical and psychological disorders.[47] Yet, despite these hardships, women may acquiesce to these appeals in order to make the painful decision to abort more meaningful, and to ensure that the resulting fetal tissue be used and not "wasted." Some ethicists have little problem with the alteration of abortion techniques designed to procure more valuable fetal organs. They see it as simply another balancing act required to save lives. Abortion advocate Dr. William Regelson has suggested that new birth control drugs like RU 486 may "permit women to time their fetal loss, making the fetus available for donations."[48]

Other ethicists and lawmakers have recommended that in order to avoid pressures on women to abort, or pressures on women to change the method and manner of abortion, there be a strict separation between procurers of fetal tissue and prospective transplant surgeons and hospitals, and that there be total anonymity between the women donating the tissue and the ultimate recipient of the fetal parts. While these efforts should reduce coercive pressure on women deciding on whether to abort, or on abortion timing, they clearly do not suffice to fully answer the problems of consent and coercion. First, up to this point, there has been virtually no effective policing of fetal organ harvesting by the federal government or any state agency. Moreover, the prospect that federal or state officials would set up a system of effectively monitoring the thousands of abortion providers nationwide appears impractical and unlikely.

Spare Parts as Art

A final objection to unlimited use of fetal parts is that it will further reinforce our already inhumane and reduced concept of the fetal body. Undoubtedly, this objection has appeal. There is something both disconcerting and tragic in the practice of using the bodies of the unborn as mere spare parts or commodities to keep an increasing number of an aging generation alive. Though falling short of prohibiting fetal parts transplantation, the Law Reform Commission of Canada in its report on human experimentation expressed this concern: "The law should never treat embryos and fetuses as mere objects. Accordingly, their commercialization or the commercialization of research on them should be strictly prohibited. . . ."[49]

Dr. Kathleen Nolan, though a supporter of fetal tissue use, has written movingly about the diminishment of family and human life inherent in the use of fetal organs and organ subparts. Nolan warns that fetal farming could badly mangle the concept of parenthood if we nurture the young not for themselves, but as medical devices for ourselves or others. She notes that "to plan the destruction of this new identity, even at early stages, is to turn one of humanity's most intimate and wonderful activities into an objectifying and pharmaceutical production mechanism."[50] There can be little disagreement with Nolan that at least ideally parents should compose rather than decompose their offspring. Reproduction to replenish fetal warehouses would represent an astonishing reduction and depletion of our common humanity.

Behind concerns about the commodification of fetuses lies the lingering and explosive issue of the legal status of the fetus. Currently, given the lack of constitutional protection for the fetus, there are few legal barriers to prevent private firms, researchers, and hospitals from utilizing fetal tissue as they see fit. For example, there is no current plan to restrict the use of fetal tissue to the treatment of serious disease. This concern was highlighted recently by the claim of Canadian researchers that some athletes in the next Olympics may be running on "baby power"—the injection or implantation of fetal tissues to enhance performance or to heal injuries. "A certain percentage of athletes will do anything to win, and fetal tissue is the next thing they'll be into," states Gerard Thorne, a well-known expert in the area of athlete "doping." Thorne's fellow researcher Phil Embleton explains how the tissue would be used: "Animal studies have shown that an injection created from fetal tissue will help an injured muscle heal much more quickly. That would be the initial use." Embleton goes on to predict that fetal tissue implants will be used in the future to stimulate an athlete's own production of testosterone and growth hormone: "Once it's shown to work at healing, the next step is to see how much it can do for a healthy body. It will follow the same pattern as steroids when they were first developed." Embleton and Thorne do not believe that ethical concerns about using fetal tissue for a nondisease purpose will stop its use for enhancing athletic performance. "When you're dealing with multi-million-dollar salaries, morals go out the window," they say.[51]

Exactly how reductionist our view of fetal parts could become is unknown. However, a recent bizarre legal case in England may be instructive. The 1989 case involved an artist, Canadian sculptor

Rick Gibson, and a London art gallery owner, Peter Sylvaire. Gibson had created a model's head, and to each of its ears had attached earrings composed of freeze-dried aborted fetuses of three to four months gestation. Gibson obtained the fetuses from a professor at a London college. He attached the fetal earrings by screwing a fitting into the fetus's skull and attaching the upper end of the fitting to the lobe of the model-head's ear. The earrings were displayed with forty other exhibits at Sylvaire's Soho gallery, The Young Unknown's Gallery. The article was described in the exhibit's catalogue as "Human Earrings."

Gibson had advertised his artwork prior to the opening of the exhibit, and several individuals had notified the authorities about the "Human Earrings." Police seized the exhibit two hours after viewing began. Both Gibson and Sylvaire were charged with causing a public nuisance and with outraging public decency. After a four-day trial in February 1989, a jury acquitted the two men by a 10 to 2 margin of the nuisance charge, but convicted them on the public decency count. Each was fined several hundred dollars. Gibson and Sylvaire appealed the jury's verdict based on the fact that the obscenity laws of England were not intended to cover this kind of artistic expression. The appeal was ultimately dismissed in July 1990. However, because the fetus had no legal protection, the court noted that only a limited number of statutory grounds were available to the court and jury to prohibit such actions.[52]

Clearly, in the absence of constitutional "personhood," some protection for the fetus is called for if it is not to become just another commodity in the human body shop. Both pro-choice and pro-life advocates should join in the search for an acceptable legal regulation before the commodification of the unborn becomes complete.

Currently, regulation of fetal use and procurement is inconsistent, and often nonexistent. In the United States, less than half of the states have any regulation on fetal research. And though most polls show that a majority of Americans oppose fetal transplantation,[53] only seven states do not allow for research with aborted fetuses.[54] Worldwide, at least five countries have guidelines on transplanting tissue from electively aborted fetuses. France, the United Kingdom, Australia, Canada, and Sweden have attempted to reduce the threats of commercialization, coercion, vivisection, and the encouragement of abortion endemic to fetal research by recommending a variety of restrictions. None of these suggested guidelines has the force of law.[55]

In March 1988, the Reagan Administration declared a moratorium on U.S. federally funded research involving the transplantation of fetal parts taken as a result of elective abortions. The controversial moratorium did permit the use of fetuses from ectopic pregnancies and from spontaneous abortions. The moratorium was inconsistent in its coverage, in that it allowed federal nontransplantation fetal research to continue. In fact, during the first three years of the moratorium, the National Institutes of Health spent more than $23.4 million in federal funds to support 295 nontransplant research projects involving use of human fetal tissue, including McCune's experiment involving transplanting fetal organs into mice. The NIH also was spending over $100,000 each year to maintain a fetal tissue bank at the University of Washington Central Laboratory for Human Embryology. This tissue bank has provided over 10,000 human embryonic and fetal specimens to hundreds of clients, including high school and college teachers for use in science classes.[56]

However, despite its inconsistencies, the moratorium had a significant public policy message for the United States and other countries around the world, one that goes well beyond abortion politics. It represented an important and unique moment in the young history of the human body shop. The moratorium was one of the few measures our society has taken to set aside time for reflection on the short- and long-term risks of the use of a new technology. Every new technology comes with benefits as well as potential problems. Too often in the past, our scientists and corporations have rushed ahead without taking time to ponder the important ethical questions.

Throughout its tenure the moratorium was consistently subjected to congressional and court challenges, but was able to survive. However, on January 22, 1993, two days into his administration (and the 20th anniversary of *Roe v. Wade*), President Clinton lifted the fetal tissue use ban by executive order. Clinton took this action without awaiting congressional action, which would have ensured at least some restrictions on fetal tissue procurement and use. The President's order signalled the start of a virtually unregulated industry in fetal parts. Clearly, the moratorium should have been continued until the problems of fetal marketing, fetal vivisection, informed consent, coercion, alteration in the method and manner of abortion, and utilitarian reductionism of fetuses have been fully considered and addressed.

———

The controversy surrounding fetal tissue use demonstrates many of the most tragic dilemmas in the field of organ transplantation. However, the commercialization of fetuses also breaks ground in another human body shop arena: the sale of the products and processes of human reproduction. Just as transplantation technology opened the door to the sale of organs, cutting-edge reproductive technologies used in the treatment of infertility are creating open markets for a variety of human reproductive material. In the next section, we will survey the rapidly growing field of "reprotech." We will see the baby factory at work, and we will describe current controversies over the manipulation and marketing of sperm, eggs, embryos, and children.

PART TWO

The Baby Factory

4

The Business of Baby-Making

MICHAEL AND DIANE are in their late thirties. Each delayed marriage in order to pursue a career, and now they are ready to settle down and have children. Married for more than a year, Michael and Diane have been unsuccessful in their attempts to conceive a child. Following many emotional discussions, and after reading positive reports in several national magazines about overcoming infertility, the couple has decided to consult one of the 250 infertility clinics that have sprouted up around the country (about two hundred in the last five years alone).

At the clinic the couple learns that Michael has a low sperm count. Treatment starts with minor surgery to see if it can be increased. After the surgery fails to improve his sperm count, Michael and Diane decide to attempt fertilization of Diane's eggs with sperm from a paid donor. The couple chooses the sperm based on the physical and psychological characteristics of the donor listed in a questionnaire. They opt for traits that are close to Michael's. The sperm, which has been preserved through freezing, has also been checked for AIDS. Diane is then given Pergonal, a powerful hormone that will cause her to produce many eggs at once. This is followed by eight unsuccessful rounds of artificial insemination with the donor sperm. Then, after several eggs are surgically removed from Diane, their doctor attempts in vitro fertilization six times. This entails uniting sperm and egg in a petri dish, with the resulting embryo being transferred to her uterus via catheter. These attempts also fail.

Their doctor explains to the increasingly frustrated couple that Diane's eggs and uterus do not seem able to function adequately, perhaps because of her age. He suggests using a healthy donor egg from a younger woman and informs them that the egg can be purchased for $2,000 from a list of young women whose characteristics and background they can review, just as they did the sperm donor. The chosen donor eggs would then be fertilized with the donor sperm, with the resulting embryo(s) to be transferred into Diane's uterus, or that of a surrogate mother, whose services could be purchased for $10,000.

When Michael and Diane discuss this possibility, they realize that creating a child from donor sperm and a donor egg, which would then be gestated in a surrogate mother, would result in a baby genetically unrelated to them and with no less than five "parents." They abandon their effort to have a child and opt for adoption. The total cost of their attempts at pregnancy is close to $40,000—part of over $2 billion spent each year by couples attempting to overcome infertility.[2]

This hypothetical scenario is a realistic one for many couples in the United States. It is estimated that infertility—defined as the inability of a couple to conceive after twelve months of intercourse—affects 2.3 million American couples (with wives aged fifteen to forty-four). This figure represents about 7.9 percent of married couples in the childbearing age group—slightly less than one in twelve.[3]

About 40 percent of couple infertility is due to fertility problems with women. Five million women in the United States—about 8.5 percent of the 57.9 million women who are between fifteen and forty-four years of age—have impaired ability to have children.[4] Slightly over 40 percent have primary infertility, never having been able to have a child. The remainder have become infertile after having one or more children.[5] Among the major causes of infertility in women are ovaries that do not release eggs properly or scarred fallopian tubes. Damage to a woman's reproductive capabilities can be caused by a number of factors, including sexually transmitted disease, infections after surgical procedures, surgical sterilization, endometriosis, unnecessary hysterectomies, cancer treatment, and damage done by contraception devices, smoking, and environmental toxins.[6]

Though infertility is often a problem thought of solely in terms of women, reports indicate that men are the primary cause of couple infertility just as often as women. Up to 40 percent of couple infertility cases are due to male fertility problems, primarily low sperm count.[7] Other problems with male partners include impotence, genetic disorders, exposure to environmental mutagens, and sexually transmitted diseases.[8]

While female and male infertility are physical and emotional problems for millions of Americans, popular descriptions of U.S. infertility are often exaggerated and misleading. The media frequently report that infertility is an "epidemic" affecting 10 million infertile couples, or a "tragic scourge" on one in six U.S. couples. These are, of course, myths. The accurate, far lower infertility levels described above have tended to remain constant over the last fifteen years. In fact, infertility rates among couples remained virtually unchanged throughout the decade of the 1980s.[9]

Though infertility has not increased, infertility treatment has. The United States is in the midst of an infertility treatment boom. Up to 1 million American couples seek infertility treatment each year, and infertility is now a multibillion-dollar-a-year business.[10] The infertility industry, which started with in vitro fertilization (IVF), now features a veritable alphabet soup of baby-making techniques: GIFT (gamete intrafallopian transfer), ZIFT (zygote intrafallopian transfer), TET (tubal embryo transfer), PZD (partial zona dissection), MESA (microsurgical epididymal sperm aspiration), DI (donor insemination), egg donation by donor, and genetic and nongenetic surrogate motherhood. (Gametes are mature sperm or eggs capable of fertilization; a zygote is the cell formed by the union of two gametes.) The term used to cover the whole range of these new techniques is "reproductive technology." Its more familiar name is "reprotech."[11]

Recently, the brave new world of reprotech has even gone Wall Street. In June 1992, one of the country's leading in vitro fertilization companies, IVF America, held a public stock offering to raise up to $19 million in exchange for 42 percent ownership of the company. IVF America has ambitious growth plans for these investment dollars: it hopes to become the McDonald's of the baby-making business, opening IVF franchises around the country.

IVF America's move, as well as the general boom in the business of reprotech, concerns many consumer advocates, health care specialists, and legislators. As noted by Oregon congressman Ron Wyden, "What you've got here is a very combustible mix, essentially no regulation, large sums of money, constantly changing and improving technology, and vulnerable couples who want to have a baby more than anything else on earth."[12] Concerns about IVF America's franchising plans are also founded in the fact that the company has already been disciplined by the Federal Trade Commission for potentially misleading advertisement of its services and seems to have a history of publicly exaggerating its success in creating babies.[13]

The bullish market for reprotech, which IVF America and others are attempting to tap, has been created by several factors. These include delayed childbearing by the Baby Boom generation, which has increased the number of childless couples with wives aged thirty-five to forty-four.[14] These older couples also tend to have more disposable income—an essential component, because reprotech is expensive and often not fully covered by insurance. Most couples also have unrealistically high expectations of infertility treatment—hopes which have been fanned by media hype. Stories with headlines such as "Miracle Baby," "Technology's Gift," or "Dream Come True" create the mistaken idea that reprotech has successfully conquered infertility. Each new advance in reproductive technology, whether in artificial insemination (AI), egg donation, in vitro fertilization, or surrogate motherhood, is heralded by the press as a key breakthrough and entices more couples to shop in the reprotech marketplace.

The use of infertility services is also being pushed by the growing number of physicians and others making their livelihoods in reprotech.[15] Currently, infertile patients obtain care from an estimated 45,600 physicians; 20,600 obstetrician-gynecologists; 17,500 general or family practitioners; 6,100 urologists; and 1,400 surgeons.[16] Finally, infertility treatments are increasing because adoption options for couples are shrinking as a result of a decreased supply of infants since the Supreme Court's legalization of abortion in 1972.

The rapid increase in treatments for infertility has raised an extraordinary number of both intimate and public concerns for individuals and couples in the United States and around the world. One concern is the psychological well-being of couples undergoing treatment. Critics point out that the emotional costs to couples can be devastating. "What happens to a man who's told, 'Your sperm don't make the grade?'" asks Annette Baran, a psychotherapist and an expert in reproductive technologies. "What does it feel like being a women with aging ova that your doctor is urging you to toss out in favor of younger, fresher donor eggs?"[17] Psychological damage is also being found in the children created through certain reproductive technologies. Children born as a result of donated sperms, eggs, or embryos, or surrogate motherhood, are confused as to their genetic heritage and often spend years attempting to find their genetic parents. Those looking for their roots in the morass of reprotech are often devastated by the discovery that they were "sold," either through sperm or egg sale, or via surrogate motherhood.

Many in the consumer and health fields are also deeply troubled by the manner in which reprotech exploits its patients. Without question, the media and medical hype surrounding reprotech hides a dismal success rate. The failure of couples like Michael and Diane is typical. It is estimated that only about 10 to 14 percent of couples who enroll in IVF programs in the United States actually succeed in achieving a live birth.[18] Negative reaction to hormone treatments and poor quality of sperm and eggs are the main culprits. Some other techniques do have slightly higher success rates: Donor egg treatments are successful 21 percent of the time; artificial insemination, 38 percent; GIFT, 23 percent; and ZIFT, 17 percent.[19] Regardless of success or failure, infertility treatment can take a tragic toll on the bodies and minds of the women undergoing the highly invasive infertility procedures. Multiple doses of powerful hormones, numerous artificial inseminations, embryo implantations, fallopian transfers of gametes and zygotes, and a variety of other procedures and surgeries are routine. One woman who had undergone numerous fertility treatments stated, "This is what hell must be like."[20] Another woman summarized her experiences in reprotech as follows: "I guess humiliation is the right way to describe what the whole process was like from start to finish."[21] Women's health advocates are increasingly concerned about the impact of these procedures on women. "We are angered that these technologies are being represented as safe, effective, and in a woman's best interest," says feminist author and activist Dr. Janice G. Raymond. "They are none of these things . . . IVF clinics exist because they are immensely profitable. They aren't proliferating out of altruistic impulses for so-called desperate infertile women."[22]

Other critics of the new baby business note the irony that while billions of dollars are spent in the often unsuccessful attempt to create babies for those who can afford the high price of reprotech, the United States is currently twenty-second in the world in infant mortality, losing thousands of babies each year to poverty, drugs, and lack of adequate health care. Certain U.S. inner city areas have a greater infant mortality rate than some of the poorest nations of the developing world.[23]

A central concern about reprotech is the social impact of the expanding free market in human reproduction. As the demand for infertility treatment increases, the industry of baby-making threatens to turn the human reproductive system itself into a factory for the human body shop. Currently, there are sixteen ways to conceive a baby. Each of them require a sperm and egg, fertilization, and a gestation site for the embryo(s). But whose egg, sperm, embryo, or womb? Is each a commodity that can be sold like any other product? The selling of sperm, eggs, and embryos, and the contracting of child-bearing all raise the specter of an economically disenfranchised "breeder class" selling their reproductive processes and genetic heritage to those who can afford the price. Economic coercion of donors to aid in the creation of children who are often their genetic offspring, but whom they will never be allowed to know, can and has created tragic consequences.

The sale of an element of reproduction is unique among all the transactions that take place in the human body shop. It is not like the sale of blood, tissues, or even organs. Implicitly or explicitly, it is the sale of a potential new person. It also represents the invasion of the market into our most intimate selves—our sexuality, our self-image, and our marriage and parenting relationships. Reprotech represents a disturbing alteration in our social and legal view of the human body and childbearing. It has thrown into doubt the very definition of parenthood. We no longer are sure what legally constitutes mother or father. We do not have a legal definition of human embryos—are they property or people? And what of sperm and eggs—should they be viewed as commodities, no different from any other? As we will see in the following chapters, our society has produced few answers to these profound and unprecedented questions.

5

If you are a healthy, intelligent male, 18 to 35 years old, with a good sperm count, we need you. Become a paid sperm donor.
    College newspaper ad for sperm donors.[1]

The Seeds of Life

SPERM: THE WORD is derived from the Greek "sperma," meaning seed. Since antiquity, sperm has been seen as the very essence of human fertility. Through ritual and art, numerous cultures have celebrated the phallus and its generative fluid.

In more recent times the role that sperm plays in reproduction has been understood scientifically. This in turn has allowed for various techniques of artificial insemination to be developed. The understanding and use of sperm may have diminished its symbolic importance, but they have allowed sperm to gain in commercial value. In fact, sperm is currently the leading commodity sold in the reprotech marketplace.

The use of sperm in artificial inseminations is not new. Spermatozoa were first "discovered" in 1677, and the first reported case of artificial insemination in animals was in 1785. The pioneer inseminator was an Italian churchman, Abbe Lazarro Spallanzini. He injected dog sperm into the vagina of a bitch, and sixty-two days later she became the mother of "three vivacious puppies," two males and one female. Spallanzini's successful experiment created a furor in the scientific world of the time. Many were concerned about the ethical implications of the technique, especially on humans, but others were excited by the prospects. In a letter to Spallanzini, French biologist Charles Bonnet wrote, "I am not so sure but that what you have just discovered may some day have extraordinary consequences for mankind."[2]

Within a few years of Spallanzini's work with dogs, the first human artificial insemination was performed. The famous English surgeon Dr. John Hunter succeeded in artificially impregnating the wife of a linen merchant with the sperm of her husband. The pregnancy resulted in a normal birth. The first artificial insemination in the United States was carried out in 1866 by Dr. J. Marion Sims. Sims reported that he had performed fifty-five artificial inseminations on six women. In each case he used the sperm of the patient's husband. Later, influenced by a religious conversion, Sims recanted his ways and abandoned artificial insemination, calling the practice "immoral."

The pioneer of artificial insemination using nonhusband "donor" sperm was Dr. Robert L. Dickinson, who practiced in the United States in the 1890s. Not surprisingly, Dickinson's work created an uproar. His use of sperm from nonhusband donors was castigated by religious leaders and others, who felt that such insemination broke the bonds of marriage and created a technological form of adultery. Despite controversy and condemnation, Dickinson's breakthrough led to the increasing use of artificial insemination with anonymous donor sperm in the early decades of the twentieth century. One donor insemination (DI) expert pays tribute to Dr. Dickinson, stating: "He was the initiator, the teacher of all of us in the field. To him belongs the greatest share of credit for the increasing practice and acceptance of donor insemination."[3]

DI has come a long way since Dickinson. The year 1953 saw the first reported pregnancy after insemination with frozen donor sperm. More recently, physicians are using extensive questionnaires to screen donors for "desirable" traits. Over 90 percent of DI providers are willing to match donors with patients on the basis of race, eye color, complexion, and height. Over 80 percent are willing to screen for ethnic or national origin, weight, body type, and hair texture. A majority of physicians offering DI will also match specifications concerning a donor's educational attainment, intelligence quotient, and religion.[4] At least one sperm bank, the Repository of Germinal Choice in Escondido, California, offers only semen samples of exceptional thinkers and athletes and became known as the "Nobel Prize Winners' Sperm Bank."[5]

You can even choose the sex of your DI child. Fourteen percent of all practitioners regularly performing DI, and 31 percent of physicians with over one hundred DI patients say that they offer "sperm separation," a technique used on sperm that allows physicians a high percentage chance of preconception gender selection.[6] In the future it is expected that advances in genetic screening will allow a physician to detect a large number of genetic traits and abnormalities contained in sperm.

The business of artificial insemination has grown along with the technology. Today, over 11,000 physicians provide artificial insemination to about 172,000 U.S. women each year. Live births are achieved in about 38 percent of cases, resulting in an estimate of 65,000 AI babies each year, about 30,000 from donor sperm. Almost half of all women involved in artificial insemination are inseminated with sperm from anonymous donors, usually medical students.[7] Paid sperm donors average $50 per donation and may give as often as two to three times a week for several years. Some students make hundreds of such donations. Sperm banks process and test the donated sperm, which is often frozen for later use. Sperm samples are then sold to clinics and physicians for as much as $200 a sample. Treatment itself is relatively expensive. In her attempt to achieve pregnancy, a woman can spend over $300 in initial consultations, examinations, and testing, and over $100 for each of several inseminations, for a total cost of over $1,000. Each year Americans spend about $165 million on artificial insemination, with insurance companies paying only about one-quarter of the sum.[8]

Despite the high visibility of the fast-growing DI industry, the commercialization of sperm donation has caused surprisingly little controversy. While the ethics of selling organs and fetal tissue and contracting for childbearing (surrogacy) are often in the news, similar concerns about sperm sale rarely surface. The ethical dormancy about DI and sperm sale is puzzling, for unlike donation of many other bodily substances and parts, donated sperm can create the irreplaceable—a child. And the impacts of DI on the donor, the couple, and the child can be significant. Unfortunately, these impacts are little understood and there is little attempt to prepare patients for them. A study done by the congressional Office of Technology Assessment (OTA) found that only 3 percent of DI practitioners discuss psychological impacts of the procedure with women receiving it, and only 1 percent discuss psychological complications for the husband or offspring.[9]

There is also no counseling for the paid donors. These men, mostly medical students in need of ready cash, generally respond to ads in college newspapers, such as the one that begins this chapter. Infertility researchers Annette Baran and Reuben Pannor, after conducting numerous interviews with paid sperm donors, created a composite portrait of the average sperm seller who answers such ads. Typically, the paid sperm donor is a student, unmarried, and emotionally incapable of understanding the long-term implications of his act on himself, the children he is helping to create, or the couples whose lives he has touched. A typical donor attitude is reported by Baran and Pannor: "Hell, man there's nothing much to it. . . . You make fifty dollars a week for a few minutes of work; you can't really call it work, because you jack yourself off for free anyway. This way you save it to get paid for it."[10] The paid donors view the inseminations as "business transactions." They see themselves as "vendors," offering a product for which there is growing demand. The anonymity promised to them by the sperm-buying clinic or doctor helps sustain the sense of detachment and lack of accountability.

However, as donors mature, get married, and have children, they begin to understand the importance of what it means to produce a child. They no longer see the sperm donation as a commercial transaction, but rather as an actual fathering. Baran and Pannor report: "The parent-child relationship awakens in ex-donors a sense of regret, concern, and fear for those children whom they fathered without any recognition of their fatherhood. Many of them see that person in that period of their lives as irresponsible and immoral."[11] A study done by Houston psychologist Patricia Mahlstedt at the Baylor College of Medicine underscores the emotional attachment that sperm donors feel for their offspring late in life. In Mahlstedt's study approximately 60 percent of ex-paid donors said they were willing to be contacted by their donation's offspring, when the children reach age eighteen. Nearly all donors were willing to fill out extensive questionnaires and have them passed on to recipient families.[12] When asked what message they would give to a child created from their sperm, 75 percent of men came up with a message immediately. "My love and thoughts are with you," wrote one donor. "Also, don't worry if you're tall and thin as a youth, because by the time you're about 20 years old, you'll have to knock the women off with a stick."[13] As donors attempt to come to terms with the results of the sale of their genetic heritage, they have additional concerns. Some donors, especially those responsible for dozens of births and who live in small towns or cities, are concerned that their offspring may unknowingly marry one another. A few such instances and near misses have already been reported.[14]

The recipient couple also is strained by DI. Women often require several inseminations of donor sperm, and donor insemination is far from a comfortable psychological experience. "I remember the first time, you do feel kind of violated, in a way," recalls one woman recipient.[15] After the birth of the child, some women fantasize about the paid donor. On woman notes, "It's funny, I have a great affection for him. . . . [My child] has so many delightful characteristics that are not mine. I think this man must be a delightful person. I really like him."[16] Husbands often report wrenching feelings of inadequacy. When his wife suggested DI, one husband recalled, "I initially said no . . . if I can't impregnate my wife I don't want any other man, even in absentia, impregnating her. . . . On top of that was the concern that his [the paid donor's] child won't look like me; he'll look like her and some other man I've never met."[17] Men also confess to overwhelming feelings of jealousy toward the paid donor.[18]

As the child grows up, the recipient couple must also decide whether, when, and how to tell the child about his or her genetic heritage. Often the couple decides to keep it a secret. In one study 85 percent of couples had not, and said they would not, tell their children that they were conceived through paid donated sperm.[19] Whether the secret is kept or exposed, the impact can be devastating, both on the couple's marriage and their relationship to the child. "These secrets are like time bombs," reports researcher Pannor. "They create a tremendous pressure on everybody."[20] Divorces based on couple tensions over DI are not unusual.[21]

It is also difficult for the children created through DI. They often become curious, even obsessed with discovering their genetic background. "Every human being has a right to their heritage," stated one DI offspring who is now looking for her father.[22] Along with the complex psychological difficulties of realizing that their fathers are not their genetic fathers, many offspring of DI are angry. They are frustrated by the wall of anonymity around their genetic fathers, and they are deeply disturbed that their births were part of a business transaction. They have feelings of being rejected and sold by their biological fathers. "They accepted money to create you," says one DI offspring. "If your own flesh and blood sold you, it's a real hard place to come from."[23] As more DI children grow up and pursue their biological roots, legal wrangles are sure to follow. "I think you are going to see a whole new generation of legal problems," states Lori Andrews, a legal expert in infertility technology, "where the children born through these arrangements are suing for information about the biological father. I've talked to a number doing extensive searches. Some are very angry, saying, 'My father sold me for $25.'"[24]

Despite its profits, and the wide range of problems it creates, DI is remarkably unregulated. The United States has a "laissez-faire" free market in sperm. Only one state, New York, is even considering forbidding payment for sperm donors.[25] There is so little official oversight of DI that no government or professional group has any idea how many sperm banks exist. In fact, since sperm banks are so little regulated and have no official reporting requirements, the real number of donor-sperm babies is not known. As noted, it is generally thought that about 30,000 DI babies are born each year, but other estimates range from 25,000 to 100,000 per year. Only fourteen states have laws or regulations governing sperm donations. Most of these laws are lax and require little more than testing sperm for HIV. The threat of sperm containing HIV and infecting the mother, and perhaps the child, has now caused the federal Food and Drug Administration (FDA) to consider regulating DI. One FDA official noted, "In previous years, we felt that it didn't need regulation. But once HIV came along . . . then FDA began to look more carefully at it."[26] But even with the threat of HIV infection, no federal regulation of donor artificial insemination has been forthcoming.

Even the issue of who is the parent of the DI child is not clear. Thirty-two states have donor insemination laws stating that the husband of the woman who is inseminated is the legal father. Seventeen of the thirty-two states say explicitly that the donor is not the legal father. But in the remaining states, the legal situation is ambiguous, particularly for single women who become mothers in this way. In certain cases, when the artificial insemination has not been conducted through a hospital or clinic but accomplished privately, sperm donors have been granted visitation with the children created by their sperm. As for disclosures about donors being available to DI children, fifteen states require doctors to file reports on DI procedures that they conduct. The laws provide that should there be good cause (for example, the need to track a potential genetic disease), the files can be opened for review.[27]

Internationally, DI has also gained general acceptance. Regulation, however, remains inconsistent. Currently, over thirty countries have legislation ensuring that the child conceived by donor insemination be legally presumed the child of the sperm recipient and her consenting spouse. Only two countries, Brazil and Sweden, have forbidden insemination by donor. Several countries limit donor insemination to married couples. The United Kingdom limits the number of times a donor can provide sperm. Certain countries, such as France and Switzerland, forbid payment to donors.[28]

———

As sperm continues to be bought and sold, processed, frozen, and resold, important questions remain unanswered. Should sperm be sold as a commodity? Can it be exchanged or legally held as property? Is DI a form of technological adultery? Do the practice and procedures of donor artificial insemination violate a sense of the respect and dignity due human procreation? Does it destroy marriages? Should children of DI have access to their genetic fathers? Should donors be allowed to communicate with their genetic offspring? Should DI be granted to any woman who requests it, regardless of marital status, criminal record, or mental state? One commentator has noted, "Our culture, our society, medicine, religion, and our morals and value system do not have the answers to this phenomenon right now."[29]

6

I think it's a bad way to make babies.
    Dr. Arthur Caplan, Director of the Center for Bioethics
    at the University of Minnesota[1]

The Price of Eggs

OVER THE LAST five years, a new body shop business has started: the selling of human eggs. The use of eggs (ova) in the treatment of infertility is far more recent than that of sperm; it was not until in vitro fertilization of ova became possible in 1978 that eggs became potential commodities for the infertility business. Once the techniques for fertilizing eggs outside the womb were developed, eggs could be donated or sold by donors and then fertilized and inserted into the wombs of women who could not produce their own eggs. With their new value in treating infertility, eggs followed sperm into the marketplace.

As a new fertility commodity, ova are in plentiful supply. Woman do not generate eggs throughout their lives. All the eggs a woman will ever produce, about 7 million, are formed during the first four months of fetal life. Even prior to birth, the process of atresia, in which eggs start their development only to degenerate and be lost, has begun. By the time an infant girl is born, only about 1 million or 2 million eggs remain. By puberty, when eggs become candidates for ovulation and fertilization, only about 300,000 remain. With one egg ovulating every month, only a few hundred of these eggs will ever have the chance to be fertilized.[2] The rest, about 299,000, are available for donation or sale. Storing available eggs is a problem. Eggs are far more fragile than sperm, and it is more difficult to freeze and preserve them. As of the late 1980s, only three births had been recorded from frozen eggs.[3]

The human egg market has become a profitable industry in recent years. "There is a lot of consumer demand," states Dr. Mark Sauer of the University of Southern California, who runs one of the nation's largest egg donor programs. "There are a lot of patients out there in their late 30s and 40s who have had a very difficult time of it."[4] Customers for the eggs are usually women whose eggs have been degraded in number or quality by age or accident, or who are postmenopausal and have no eggs. The eggs of donors are the last hopes many of these women have of bearing babies—they have a 20 percent to 25 percent chance that implantation with a fertilized donor egg will result in a live birth. The added bonus for older women is that donor eggs from younger women promise to have fewer chromosomal birth defects.

In response to this demand, over sixty-five medical centers around the country are offering donor egg programs for infertile women. If a woman wants to become impregnated with an egg from a donor that has been fertilized by her husband or other chosen party, these clinics will help. They will fertilize the eggs in the laboratory and then implant them in the infertile women's womb for gestation. In 1990, clinics reported almost 550 egg transfers.[5] In addition to the $12,000 fee, there is one catch: almost all of these centers require that the infertile women provide her own egg donors.

Not surprisingly, many women cannot bring themselves to ask a friend or relative to provide eggs. Even for those women who do have the gumption to ask, egg donors are not easy to come by. "It is very difficult to go out and find a donor," reports Dr. Sauer.[6] Dr. Zev Rosenwacks, who runs an egg donor program at New York Hospital–Cornell Medical Center concurs. "I don't know many women who want to become donors," he said.[7]

The lack of enthusiasm for donating eggs is not surprising. Human eggs, unlike sperm, are hard to collect, and collection can be hazardous to the donor. Prior to donating an egg, a woman must be injected with hormones to overstimulate the ovaries to produce large numbers of eggs, and they must also undergo frequent blood tests and ultrasound scans to determine when the eggs are ready to be fertilized. Finally, anesthesia is required when the eggs are to be removed.

The removal of the eggs is dangerous in itself. "Since ovaries are internal, the eggs have to be sucked out with a needle," notes Dr. Sauer. "There is the risk of hemorrhage and infection, and therefore a risk of damage and potential infertility." The hormones given to egg donors can also have side effects, including ovarian cysts. "Women can injure their reproductive organs going through these cycles," states Richard Dickey, medical director of the Fertility Institute of New Orleans. "I guarantee that some of these donors are going to have reproductive problems in the future."[8]

As the intrusive nature of the procedures involved in egg stimulation and retrieval make donors scarce, there is a growing trend to pay fees to women in order to entice them to donate eggs. Clinics are advertising for egg donors in magazines and newspapers across the country. One ad, from IVF New Jersey, under the bold heading "Earn $2,000," urges readers to "Help infertile couples realize their dreams."[9] An ad placed in *The Broadside*, the newspaper of George Mason University in Virginia, reads: "Healthy Women Wanted as Egg Donors. Help Infertile Couples. Confidentiality Ensured."[10] Currently, ten clinics in the United States supply women clients with lists of healthy young women who have agreed to provide eggs for about $2,000 per removal.

Under the guidelines of the American Fertility Society (AFS), women are not supposed to be paid for the eggs themselves, but rather for "expenses, time, risk, and inconvenience associated with the donation." Ultimately, however, there is little real difference between paying for the eggs or for the process of their retrieval. "It's almost a matter of semantics," admits Joyce Zeitz, public relations coordinator for AFS. Legal scholar George Annas sees the regulation as a ruse: "Of course, you're not really buying a woman's inconvenience. It's a bogus argument that you're not actually selling these eggs. Clearly, donors are selling their reproductive capacity. And if you can sell your egg, then why shouldn't you sell your child too?"[11]

Most often the sellers of eggs are college students who need the extra cash and welcome the free physical exam offered as part of most paid egg donor programs. Some clinics screen paid donors to eliminate women with poor medical histories and to find donors who match the personal appearance or temperament needs of clients. Psychologist Georgia Wirkin, who screens for the Mount Sinai Hospital's egg donor program, comments, "One donor's hobbies might be skydiving and mountain climbing. You wouldn't match her egg with a couple whose idea of a good time is staying at home and doing crossword puzzles."[12] The psychologist does not explain how taste in hobbies will be genetically transmitted through the egg.

Paying donors for eggs troubles many in the bioethics field. They are critical of a practice that allows the rich to buy the irreplaceable genetic material of the poor, and which coerces the paid donor into submitting to physical risk. And, without question, money is coercive. Donors readily admit that without the money they would not give up their eggs. "If I were not paid, I definitely would not do it for strangers," noted one donor, a nursing student at a college near Washington, D.C.[13] Another donor, a student who has kept her donor status from her "deeply religious" family, confessed, "I needed the money to continue my education."[14] Yet another student donor, an economics major at George Mason University (who responded to the ad in *The Broadside*), also hasn't told her parents. "I don't want them to think, 'Gee we can't afford to put [our daughter] through graduate school, so she's having to sell her body parts.' "[15]

Induced by the fee to sign up with a donor program, many egg donors then face physical exploitation. Katie, a thirty-three-year-old woman in Brooklyn, New York, was the focus of an exposé of the Mount Sinai egg donor clinic published in New York's *Newsday*.[16] Katie was having trouble making ends meet and was told by a friend about the Mount Sinai egg donor program. Hoping to make some easy money, and help an infertile woman at the same time, Katie went to the Center for Reproductive Services at Mount Sinai and signed up for the program. At that time they were paying $1,500 to egg donors per retrieval. Katie, without being informed about the risks involved or signing a consent form, began the Mount Sinai four-week program required prior to egg donation. The program involved daily self-injections with massive doses of the hormones Pergonal and Metradin to stimulate egg activity. It also required frequent blood tests, pelvic examinations, and ultrasounds. For three weeks all appeared normal.

However, in the fourth week, doctors at the clinic discovered that Katie had developed an ovarian cyst. As noted, cysts are a common side effect of large doses of Pergonal. The cyst blocked the ovary that was to provide the twelve to fifteen eggs that the clinic was planning to harvest from Katie and then implant in other women. After the cyst was discovered, Katie was dropped from the program and paid nothing. She was crestfallen, "I did everything they asked of me, and they made me feel like a bad laboratory rat who should be tossed on the trash heap."[17] Katie protested to the Center's director, Dr. Daniel Navot, who insisted that Katie had not undergone the hardest part of the program—surgical removal of the eggs under general anesthesia—and therefore should not be paid. Later the clinic offered Katie $350, and eventually she received $1,000. Meanwhile, reporters discovered that the Mount Sinai clinic itself was not even licensed by the state as an egg donation clinic.[18]

Arguments against paid ova donation go beyond economic coercion. One concern is that egg donors may claim motherhood over their genetic offspring after their birth. Moreover, as with sperm donation, children born from a paid egg donation may face psychological harm if they learn that their genetic inheritance was not donated out of care or love, but rather sold for $2,000. The number of unanswered ethical questions about egg donation lead some ethicists like Dr. Arthur Caplan, Director of the Center for Bioethics at the University of Minnesota, to conclude that paid egg donation is not worth its risks: "I think it's a bad way to make babies," says Caplan.

Others tend to dismiss the many concerns over paid egg donation. Dr. John Fletcher, bioethicist at the University of Virginia, is sanguine about egg use: "I think it's a good thing to donate eggs to have a baby. The more wanted babies there are, the more happiness there is."[19] Supporters of paid egg donation note that paid sperm donation has caused little stir, and that egg donation is a similar procedure. This line of argument appears flawed. As has been noted, paid sperm donation has not proven itself to be problem-free, and egg donation has features that make it even more exploitative than sperm donation. At $2,000 per donation, ova donors are more subject to outright economic coercion than are sperm donors. Moreover, egg donation is far more intrusive and dangerous for the donor than sperm donation. The very intrusiveness of the ova retrieval process results in donors being genetic parents of fewer children than sperm donors—children for whom they have physically "suffered." Ethicists fear that this may lead donors to seek out their genetic offspring more frequently than have sperm donors.

Despite these concerns, clinics involved in egg donation are essentially unregulated. The laissez-faire market approach used in sperm sale is being extended into egg sale. Scandals like the reported incident at Mount Sinai are believed to be widespread. Yet few states have attempted to prohibit egg sale. "[E]gg donation is so new that only a few states have faced it at all," states Lori Andrews of the American Bar Foundation. Only Louisiana has banned the sale of ova, and only Oklahoma has a law explicitly allowing compensation.[20]

Internationally, there appears to be more alarm over the exploitation and degradation of women and childbearing involved in ova sale. Many countries have taken regulatory action on egg donation. Some, including Australia, Germany, Israel, and Sweden, have forbidden the sale of ova by donors.[21]

The selling of sperm and egg lead inevitably to questions about the status of that which they create together: the embryo. Is the embryo qualitatively different from the separate elements that produce it? Is the embryo also property, or a commodity, that can be sold? These questions bring up cutting-edge controversies surrounding the moment when human life begins, and the essence of what it means to be human.

7

Once there is a union of sperm and egg, what you have is a living entity with a full and unique genetic character. There shouldn't be a moral coarsening of appreciation for life that allows us to think of it as nothing.
Robert Royal[1]

Embryo Imbroglio

IT BEGAN ON July 25, 1978, when British researchers Patrick Steptoe and Robert Edwards culminated years of experimentation by delivering through cesarean section a five-pound, twelve-ounce baby—Louise Joy Brown, the world's first test-tube baby to be carried to term. At the time the birth was called a "miracle." In retrospect, the process appears relatively straightforward. An egg and sperm were united in a petri dish, the in vitro-fertilized (IVF) embryo was transferred into a waiting mother, and after nine months a baby was born. IVF quickly caught on: Two years after the British success, the first IVF baby was born in Australia; and a few months later, the first U.S. test-tube birth took place. To date, an estimated 20,000 test-tube babies have been born around the world, about one-third of these in the United States.[2]

Reproduction technology has evolved rapidly since Louise Brown's birth. Today, new techniques allow for embryos to be fertilized in one woman, then "flushed" and implanted in an infertile patient. The first "flushing" of an embryo took place in 1983.[3] This technique is now routine. Moreover, whether created in a test tube or flushed from a woman, embryos can now be frozen for later use, by a process called cryopreservation. Freezing embryos and then thawing them for implantation has been a common practice in the cattle industry for two decades. But it was not until 1984 that Australian researchers reported the first successful human birth from a frozen embryo. Two years later, a U.S. baby was born from a frozen and thawed embryo. As of 1990, over 3,300 frozen embryo transfers had taken place in the United States, resulting in approximately 350 live births. Over 23,500 embryos are now in frozen storage.[4] Thousands of new embryos are being frozen each year.

In many cases advances in embryo manipulation have proved too successful. Researchers routinely implant multiple embryos into a woman's uterus, only to have several "catch." About 25 percent of embryo transfer births involve twins or triplets.[5] However, too many growing embryos in one uterus threaten the health of one another. A technology had to be developed to destroy unwanted "successes." To meet this need, doctors have developed a technique called, somewhat euphemistically, "selective reduction of pregnancy." After successful implantation of several in vitro–fertilized embryos into a woman's uterus, a physician injects a lethal chemical substance into one or more of the developing embryos in order to improve the chances that the remaining embryos will survive. The destroyed embryos are then absorbed by the body.

New techniques in embryo manipulation are creating possibilities and ethical dilemmas unimaginable in 1978. As we will see in the final chapter of this section, geneticists are now developing sophisticated genetic screening technology that allows them to diagnose IVF embryos for a number of genetic traits prior to their implantation. With the aid of these new genetic diagnostic tools, "defective" embryos can be identified and destroyed, and desirable ones implanted.

The business of embryo transfers has also charted new and controversial ground. In 1986, a "surrogate" mother was successfully solicited by a broker to sign the first contract for "gestational" services. She was paid $10,000 to be implanted with a client couple's in vitro–fertilized embryo and carry the couple's child to birth.

Over the last two decades, no area of reprotech has advanced more quickly than embryo manipulation. As a result, the embryo has become a valuable and malleable commodity in the human body shop. Currently, embryos—by the tens of thousands—are being flushed, frozen, implanted, discarded, destroyed by lethal injection, donated, gestated for pay, and genetically screened.

While the technology and business of embryo manipulation moves inexorably forward, profound questions about the embryo's moral and legal status remain unanswered. Do embryos have intrinsic worth as living entities? Should the embryo be viewed as property? What should become of frozen embryos in the case of the death of one or both spouses, or in the case of divorce or disagreement?

Answers to these questions are hard to come by. As of 1990, only five states had passed laws specifically dealing with in vitro fertilization.[6] Of them, Louisiana is alone in recognizing the embryo as a "juridical person," giving it the right to inherit property once it is born. The state also prohibits the destruction of frozen, in vitro-fertilized embryos.[7] Pennsylvania requires detailed reporting of IVF use of embryos.[8] Most states do not specifically prohibit the sale of human embryos, and only six states have laws that can even be interpreted as forbidding the sale of a fetus or embryo. Florida's law is the most direct: "No person shall knowingly advertise or offer to purchase or sell, or purchase, sell or otherwise transfer, any human embryo for valuable consideration."[9] Eight states forbid donating embryos for research, though exceptions are made for in vitro-fertilized eggs.[10] No state has legislated on ownership of frozen embryos in cases where couples disagree, divorce, or die. The vast majority of states have not addressed whether embryos are property, people, or something in between. "It's a real hodgepodge of laws," says Joyce Zeitz of the American Fertility Society.[11]

Internationally, the situation is equally confused. While seventeen nations have some legislation regarding research on embryos or in vitro fertilization, only one, Switzerland, has specifically forbidden commercialization of embryos. No country has specifically addressed the issue of ownership of embryos.[12]

Without legislative oversight or guidance, infertility entrepreneurs and reproductive technologists continue to expand their practices but are themselves confused. "I'm concerned with legal and ethical issues," says infertility specialist Dr. Mark Sauer. "But I don't have any answer to these problems. I don't think anyone does."[13] "Certainly, we need to regard fertilized eggs, embryos, as something different than a hamster," notes University of Tennessee geneticist Sherman Elias. "But, we can't treat them the same as we treat a child."[14] As lawmakers and practitioners settle for an unregulated approach to embryo manipulation, local courtrooms have become the principal forum in which the profound issues surrounding the embryo's legal and moral status are disputed. Each new court decision becomes the flash point in the historic debate over the definition of human life and the limits of property.

Frozen by Law

The public was first alerted to troubling questions regarding the legal status of frozen embryos when the media reported the unusual circumstances surrounding the tragic deaths of Mario and Elsa Rios. In 1981, the Rioses, residents of Los Angeles, were struggling in their attempt to have a child of their own. Their plight finally led them to a well-known IVF program in Melbourne, Australia. At the clinic, Mr. Rios, who was fifty at the time, was found to be infertile, and the couple agreed to have three eggs from Mrs. Rios, who was thirty-seven, fertilized with sperm from an anonymous local donor. One of the three embryos was transferred into Elsa Rios's uterus, and the other two were frozen. Unfortunately, Mrs. Rios suffered a miscarriage shortly after the implantation of the embryo. The couple elected to put off any implantation of the two frozen embryos until a later time.

Before they could return to Australia to use the frozen embryos, Elsa and her husband were killed in a plane crash in Chile. Their deaths "orphaned" the two frozen embryos in the Australian medical center. The well-to-do couple left no will, and legal confusion reigned. Were the frozen embryos "persons" who had rights in the estate? Could a surrogate mother allow herself to be implanted with either or both embryos and then make a claim on the estate for her child and herself? There were other heirs besides the frozen embryos. Mr. Rios had a son by a previous marriage who had a claim to the estate, and Mrs. Rios's mother claimed her daughter's share of the estate. These questions were resolved in 1987, when the Rios's estate was finally settled. The court decided, under relevant California law, that Mrs. Rios's mother was the sole heir and that the embryos had no rights to the estate. But issues could resurface if the embryos were eventually thawed, implanted, and brought to term—though given the time that has passed, the survival chances of the embryos would be minimal.[15]

A similar, more recent embryo imbroglio, the *Davis* case, involved the divorce of a couple who had created frozen embryos. The case gained national attention and brought to the fore fundamental questions about the nature of the embryo as property. Mary Sue and Junior Lewis Davis were residents of Tennessee and had been married for nine years prior to their breakup in 1989. The Davises originally met when both were serving with the United States Army in Germany. They had married young, he at twenty-one and she at nineteen. When they returned to civilian life, Junior Lewis was hired by his local town housing authority as an electrician and refrigeration technician; Mary Sue became a sales representative for a boat company. Both wanted very much to have a family, but problems quickly developed. Over the first four years of the marriage, Mary Sue had five traumatic and difficult tubal pregnancies. After the fifth, she concluded that she simply could not endure further natural attempts at childbearing. Like so many other couples frustrated with the contingencies of natural childbirth, the Davises decided to consult a fertility specialist. In 1985, they began treatment under Dr. Irving Ray King, who had recently opened the Fertility Center of East Tennessee. Mrs. Davis then underwent six IVF attempts using her own eggs and the sperm of her husband. No embryo freezing was employed. Each implantation followed the familiar IVF pattern—hormones to stimulate egg production and the reproductive system, extraction of several eggs, insemination in vitro of the eggs, implantation of the newly fertilized embryos, and then weeks of anxious waiting to determine if in utero pregnancy had actually occurred. As in so many other cases, all six IVF attempts failed. Each attempt cost the Davises $4,000 to $6,000. Discouraged, the couple left the program.[16]

In the fall of 1988, Mrs. Davis—still eager to have children—learned of the new embryo freezing program that Dr. King had initiated at his clinic. Mary Sue discussed the new technique with her husband, and they reentered King's program in order to try embryo freezing. In December 1988, nine ova were surgically extracted from Mrs. Davis, all of them inseminated with Mr. Davis's sperm. Two of the fertilized ova were implanted in Mary Sue, the other seven were cryopreserved for future implantation purposes. Neither of the two implanted embryos resulted in a pregnancy. Before the Davises could use the seven frozen embryos, they had filed for divorce.[17]

When Junior and Mary Sue first considered divorce, they agreed that their separation would be amicable. She would get the car, he would get the house, and they would split the furniture. Their modest incomes and lack of children allowed both to think that the marriage would end peacefully. It was not to be. In fact, *Davis v. Davis* became one of the most unusual, embittered, and publicized custody cases in recent memory, and the first case in U.S. history to decide the custody, ownership, and legal status of frozen embryos.

The legal fireworks started in February of 1989. Junior filed a complaint requesting that the court (1) give him joint custody of the seven frozen embryos in Dr. King's clinic; (2) prohibit Mrs. Davis or any other woman from gestating the embryos without his permission; and (3) if neither of the above, then to name Mrs. Davis as the only suitable party for implantation with the embryos. Although Junior did not want the embryos destroyed, he preferred destruction of the embryos to implantation in a stranger. As for his plea for joint custody, he insisted that he and his wife should jointly decide how the embryos should be utilized. Mr. Davis requested that until the parties agreed, the embryos should remain in their frozen state. Junior vehemently opposed sole custody and use of the embryos by his wife. Should the court so decide, he stated that he would be "raped of my reproductive rights. . . ."[18] He argued that her use of the embryos without his consent forced unwanted parenthood on him.

Mary Sue, on the other hand, testified that she felt attached to the embryos, viewed them as children, and wished to have her own children through their implantation. She requested that the court award her sole custody of the embryos for that purpose. In the event she could not utilize the embryos, she testified that she would not foreclose the possibility of donating them for use by another infertile couple. She further testified that her husband had agreed, as late as March 1989—a month after their separation—to being a father to the frozen embryos should they be implanted and brought to term. According to Mary Sue, the couple had amicably discussed Junior's visitation with the child or children born from the frozen embryos. She said she could not understand her estranged husband's change of mind.[19]

Hearing the case was W. Dale Young, Circuit Judge, Fifth Judicial District of Tennessee. Judge Young seemed to enjoy the novelty of the issues presented by the case. At the outset he established the broad scope of the issues he felt that the court needed to address if the custody issue was to be resolved: "Are the embryos human?" asked the judge. "Are the embryos beings? Are the embryos property that may become human beings?"[20] The court heard from five experts on these issues during the trial. The two leading witnesses were Texas University law professor John A. Robertson, a well-known writer on bioethical and legal problems and a member of the American Fertility Society's (AFS) Ethics Committee; and internationally known geneticist and author Jerome Lejeune, who was a key figure in the discovery of the genetic cause of Down's syndrome. Robertson offered testimony intended to support the position of Mr. Davis. Robertson described the frozen embryos as "pre-embryos," a term used by the AFS to describe the embryo until fourteen days after it is fertilized. Robertson's view was that a pre-embryo is an entity composed of a group of undifferentiated cells that have no organs or nervous system. He testified that at about ten to fourteen days, the pre-embryo attaches itself to the uterine wall and begins development. Prior to that time, according to Robertson, it is "not clear" that a human pre-embryo is a unique individual. Given this view, Robertson took the position that, legally, the pre-embryo is not a child and does not have the legal protection of a person. As such, he argued, the sole decision-makers over a pre-embryo's fate should be its biological mother and father. The state has no interest in the matter because it has no "person" to protect. In the event that the couple could not agree on how the embryos should be used, Robertson recommended that the pre-embryos should be allowed to "die a passive death."[21]

Lejeune strongly disagreed with Robertson. Rejecting the pre-embryo concept, he asserted that even the single cell created by fertilization is a "tiny human": "[A]t the very beginning of life the genetic information and the molecular structure of the egg, the spirit and the matter, the soul and the body must be tightly intricated because it is the beginning of the new marvel that we call human. . . ."[22] Lejeune argued that the embryos were not marital property, but rather equivalent to children, and that custody should go to the party that would wish to preserve those children—in this case Mrs. Davis. Lejeune asserted that "the early human beings [embryos] . . . are not spare parts which we could take at random, they are not experimental material that we could throw away after using it, they are not commodities we should freeze and defreeze at our own will, they are not property. . . ."[23]

Judge Young issued his opinion on September 21, 1989. His findings of fact and his legal conclusions followed closely the reasoning of Lejeune. He held that "From fertilization, the cells of a human embryo are differentiated, unique and specialized to the highest degree of distinction." Thus the court decided that "human embryos are not property." The court quoted Vice President Al Gore on the critical body as property issue: "I disagree that there's just a sliding scale of continuum with property at one point along the spectrum and human beings at another. I think there's a sharp distinction between something that is property and something that is not property. . . ."

The court also decided that "human life begins at conception." He ordered that temporary custody of the seven cryopreserved human embryos be given to Mrs. Davis for the purpose of implantation. He reserved questions of support, visitation, and final custody of the children created through the implantation until such time as the children were born.[24] Mr. Davis quickly appealed.

In the months during which the appeal was pending, the case took an ironic turn. Both parties remarried. Mary Sue, now Mrs. Stowe, moved to Florida. Their new marriages led both Junior and Mary Sue to change their original positions on the frozen embryos. Mrs. Stowe no longer wanted the embryos for herself. "If I'm going to have children, it's going to be with my new husband," she stated in a press interview. "I still believe that the embryos are life and should be given the chance to live and be born, maybe to an anonymous couple."[25] This change of heart prompted a judge in the Tennessee appeals court hearing the case to accuse Mrs. Stowe of being more concerned with prevailing over her ex-husband than with the survival of the embryos.[26] But Junior had also revised his position. He now wanted custody of the embryos. His new wife was incapable of having children. "If the court rules that these things are to be implanted, then I want to be their father," Junior stated at an appeals court hearing. Davis said he was prepared to hire a surrogate mother to gestate the embryos if necessary.[27]

Adding to the legal mayhem, a private attorney in Tennessee, R. D. Hash, asked the trial court to appoint him the embryos' legal guardian. "Since the judge has ruled that the embryos are life or persons, then somebody needs to represent their best interests," Hash stated. "It's unclear to me if either parent really wants them, so they are truly orphans. Medical personnel say the shelf life of embryos is two years, so if they are to have a chance to survive, somebody needs to act quickly."[28] Hash eventually submitted a "friend of the court" brief that outlined his views to the appeals court.

The couple's change of positions on the frozen embryos did not make the decision any easier for the three-member panel of the Tennessee Court of Appeals reviewing the case. "This case is really a mess," exclaimed Judge Hershall Frank, one of the judges hearing the appeal.[29] Yet on September 13, 1990, almost one year after the trial court opinion, the appeals court ruled. In an opinion written by Judge Frank, the court reversed the lower court ruling. They noted that the lower court's view of the embryo or fetus as "person" was in opposition to the Supreme Court's holding in *Roe v. Wade*. Further, the court ruled that allowing Mary Sue to implant the frozen embryos without Junior's permission violated his constitutional rights. The court strongly implied that the embryos were property, joint property of the couple, noting that "[j]ointly the parties share an interest in the seven fertilized ova." The court vested "Mary Sue and Junior with joint control of the fertilized ova and with an equal voice over their disposition."[30]

Mary Sue then appealed the case to the Tennessee Supreme Court. On June 1, 1992, the court ruled. In its decision the court noted that it had "no case law to guide us" and "no statutory authority or common law" that would help.[31] The court, in its forty-page decision, ultimately held for Junior. It stated that, generally, no individual can be forced to be a parent and that "ordinarily, the party wishing to avoid procreation should prevail, assuming that the other party has a reasonable possibility of achieving parenthood by means other than the pre-embryos in question."[32] Junior, under the decision, will have a veto power on any implantation of the embryos. In its findings the state supreme court essentially agreed with the appeals court. However, there was one important exception. The supreme court criticized the appeals court for viewing the embryos as "property." The supreme court attempted to find a middle ground on the controversy: "We conclude that pre-embryos are not, strictly speaking, either 'persons' or 'property,' but occupy an interim category that entitles them to special respect because of their potential for human life."[33] As a result of its decision, the court instructed Dr. King's Knoxville Fertility Clinic to "follow its normal procedure in dealing with unused embryos," as long as that did not involve the implantation of the embryos in another woman against Junior's wishes. A few days after the court's decision, Dr. King informed the court that the clinic's normal procedure was to donate the embryos to couples for implantation. In that the clinic's donation policy conflicts with the court opinion that Junior cannot be forced to become a parent, the fate of the embryos remains in legal limbo. On February 22, 1993, the United States Supreme Court officially declined to enter the embryo fray. The justices, without comment, let the Tennessee Supreme Court's holding stand, refusing to hear an appeal of the decision by Ms. Stowe.

In contrast to the Tennessee Supreme Court's middle position, at least one court has treated embryos as if they were commodities. In a federal court case filed in Virginia, a couple sought to have their frozen embryos transferred from a clinic in Virginia to one in California. The Virginia clinic refused, stating that the contract they signed with the couple did not allow such a transfer. The case, *York v. Jones*, decided in 1989, was the first to address a conflict between patients and clinics over the issue of ownership and control of frozen embryos. Referring to frozen embryos as "pre-zygotes," the court held that the pre-zygotes were the "property" of the couple to do with as they saw fit. The welfare of the frozen embryos in such a transfer, or the interest of the clinic in their preservation, were not held to have legal significance.[34] The case has been influential. It was cited as a key precedent by the appeals court that decided the *Davis* case.

The courts have only begun to define the legal status of embryos. However, consideration of the *Davis* and *York* cases indicates that embryos could join sperm and ova as full-fledged commodities in the body shop. This unfortunate further reduction of life to commodity would undoubtedly cause more controversy. An embryo is not, after all, simply an element of reproduction, as are sperm and ova; it is the beginning of life itself. For many, the embryo's "human" essence would make its commodification abhorrent. Yet despite the compelling view that embryos deserve respect as at least one form of human life, their current cavalier treatment, manipulation, and destruction in clinics continues without great public outrage or legislative intervention. While Congress has forbidden the sale of fetal parts, it is unlikely that they will do the same for embryos, due in part to the view that so-called pre-embryos or pre-zygotes are merely "masses of cells." All in all, it appears likely that as reprotech advances, we will soon see our first headlines announcing the first sale of an embryo, and perhaps even the first patenting of a human embryo for research use.

As society awaits the denouement of the embryo debate, the commodificaton of reproduction continues. By 1986, the *Baby M* case fought in New Jersey focused international attention on the ultimate in the commercialization of reproduction: contracting for childbearing.

8

There are in a civilized society, some things that money cannot buy.
    Chief Justice Warren Wilentz, New Jersey Supreme
    Court decision in the Baby M case

Baby-Selling, Pure and Simple

THE 1988 CLASSIFIED ad, though small, had to be among the most striking ever published in *USA Today:*

SURROGATE MOTHER NEEDED
to legally carry loving infertile couple's child.
$10,000 + Expenses paid.
CONFIDENTIAL. Blue or green eyes,
5'2"–5'8" preferred. Call collect.[1]

The ad was placed by the Surrogate Mother Program of New York City and its director, Dr. Betsy Aigen. Aigen is one of dozens of baby brokers who have, over the last decade, induced financially stressed women across the country to rent their wombs and bear children for clients who are willing to pay high prices for a child.[2]

From classified ads in dozens of national and international newspapers to fliers placed on parked cars in major cities, the message is clear: Surrogate mothers wanted. However, as the *USA Today* ad demonstrates, not just any surrogate will do. Broker's solicitations like Aigen's often advertise for surrogates with specific characteristics like blue eyes or moderate height, traits that make them and their potential offspring more attractive to selective, well-to-do clients. An ad in a Boston newspaper promised $50,000, five times the normal amount, for a surrogate mother who could fulfill the description "aged 22–25, tall, trim, intelligent and stable."[3]

U.S. baby brokers have arranged the birth of over four thousand babies through commercial contract, charging customers between $30,000 and $45,000 per child.[4] While the sale of these children has barely made a dent in the reported 2.3 million cases of infertile couples in this society, it has been profitable for many of the brokers, bringing in close to $40 million to the baby sellers.[5] (This multimillion dollar figure does not include amounts paid to brokers by the many clients who never receive a child.) In the hands of these new entrepreneurs, childbearing, one of the most important and revered of all human activities, is fast becoming a profitable and highly visible commercial business.

If the new industry of baby production is allowed to continue, thousands of women could be used each year as "breeding stock" to gestate babies for clients. As such, the new baby business could represent a unique form of bioslavery over women. Once women, enticed by a fee, sign contracts to produce babies for customers, they are artificially inseminated as many times as is necessary to induce pregnancy. The mothers are then placed in commercial servitude twenty-four hours a day for 270 days. Surrogacy agreements routinely require that the prospective mother submit to massive doses of fertility drugs, hormone injections, amniocentesis, and an array of genetic probes and tests at the discretion of the client. The agreements often stipulate that the mother agree to abort the fetus on demand if and when the client desires to terminate the "service." These contracts also have written provisions that make the mother liable for all "risks" that are incidental to conception, pregnancy, and childbirth, including all pregnancy-induced diseases, any postpartum complications, and even death.[6] The economically disenfranchised women who sign surrogacy contracts are often unaware of their legal rights and unable to afford attorneys.

In return for their servitude, the mothers who have signed surrogacy agreements are generally paid $10,000. Most often this payment is made only after the product, the baby, is delivered to the customer. Under certain contracts the contract mother receives only $1,000 if the baby is stillborn. No product, no payment.[7]

It is not only the mothers who are exploited by commercialized childbearing. The final victims in the surrogacy process are the babies. Their lives are negotiated and contracted prior to conception. The loving bond between mother and child is severed and replaced by the calculations of lawyers and the restrictive provisions of a contract. Under contract law, the child is a commodity whose status becomes indistinguishable from manufactured goods. The ultimate psychological effects on the children born under these contracts is unknown but potentially devastating.

The children born pursuant to contract childbearing remain the most troubling and publicly visible commodity in the human body shop. The selling of children—and the transformation of childbirth into a commercial service—is seen by many as one of the greatest threats to human dignity in our time. As noted by feminist author Katha Pollitt, "Surrogacy degrades women by devaluing pregnancy and childbirth; it degrades children by commercializing their creation; it degrades the poor by offering them a devil's bargain at bargain prices."[8]

Selling Lives

Throughout the tumultuous thirteen-year history of contract childbearing in the United States, media attention has focused on surrogate mothers like Mary Beth Whitehead Gould who have been caught up in legal battles to gain custody of their children. Unfortunately, little attention has been paid to the real force pushing the sale of motherhood—the baby brokers themselves. The brokers have generally been able to keep a low profile and avoid exposure of their practices. This anonymity is surprising, since the history of baby brokerage is rife with exploitation, fraud, gross negligence, deceit, and even murder.

As of April 1992, there were approximately twenty-nine brokers in the baby-selling business in the United States.[9] Of the active brokers, five were lawyers, two were social workers, two were housewives; others included a K-Mart operator and several people without known expertise or employment. A few of the brokers operated out of slick urban offices, others worked out of their homes, and at least one sold a "surrogate mother kit" as a traveling salesperson.[10]

Baby vendors are not licensed, and they obey no state or federal rules. No government agency keeps records on the number of surrogate contracts and births, or the number of miscarriages, sexually transmitted diseases, hormone injections, abortions, abandoned children, ill babies, or bungled inseminations attributable to these baby sellers. Couples and other clients who are shopping for surrogate mothers, and women seeking to become surrogate mothers, have no reliable data from which to judge the competence, performance, or honesty of surrogate brokers.

The toll caused by this ignorance has been high. At least fifty-five surrogate mothers have filed lawsuits and complaints charging brokers with abuse and intimidation. Many others have suffered in silence because they lacked the finances to undergo lengthy and grueling court fights. Additionally, at least twenty-three cases and forty-three formal complaints to state officials have been filed alleging that brokers have defrauded clients or surrogates or have been negligent in screening surrogate mothers. Several infants born of surrogate arrangements have been abandoned because they were the wrong sex; other babies have become the focus of lengthy highly publicized legal battles; still others have been left in legal limbo because they were born handicapped or ill. Yet many of the brokers, including those most involved in this continuing human tragedy, have stayed in business and have kept on raking in the profits.[11]

Take the case of Kathryn Wycoff. In 1989, Wycoff was running a baby business out of her home in San Clemente, California. California is the number one state for baby brokers, hosting at least thirteen such businesses over the last decade. Wycoff relocated in California in 1983 after she closed her Columbus, Ohio, surrogacy business. Described as an individual who "exudes the casual confidence of success," Wycoff puts her annual income from surrogacy at $50,000 and claims to have arranged fifty-two births at $30,000 each. Wycoff has boasted that she deals only with "quality surrogates."[12] However, with no professional board or government agency to consult, Wycoff's clients have no way to assess her past and the truth of her various professional claims. This is unfortunate, for Wycoff's past is reported to be riddled with inconsistencies and scandal.[13]

For example, Wycoff claims that she herself was a surrogate mother. Yet, as reported by authors Rebecca Powers and Sheila Gruber Belloli, her ex-husband and the broker she claims she worked for deny the story. Wycoff also claims that she has an education degree from Otterbein University in Ohio. The university, however, told reporters that they had no record of her attendance. Even more troubling is that Wycoff left Ohio while her surrogacy office was being targeted by a state probe requested by the county prosecutor into whether she was violating child placement laws.[14] The investigation was dropped when Wycoff moved west. Wycoff's move also spared her further involvement in one of the most convoluted and tragic legal tangles associated with surrogacy in the United States.

In 1985, as part of Wycoff's program, surrogate mother Lee Stotski gave birth to a contract child, a daughter named Tessa, for client Richard Reams and his wife, Beverly Seymour. A short time later, Reams and his wife separated. As part of the divorce and custody proceedings, the court ordered blood tests of the parents. The tests revealed that Reams was not the father of the contract child. Further investigation revealed that Stotski had not been inseminated with the sperm of Reams but with that of a coworker. Despite the lack of genetic connection, both of the divorcing parents wanted custody of Tessa. Stotski also wanted custody of her child, to protect her from the trauma of the divorcing couple. Soon a three-way custody battle for Tessa emerged between Stotski, Reams, and his ex-wife. Stotski eventually "reluctantly" dropped out of the court battle due to stress and financial problems.[15]

Wycoff's complicity in the various misrepresentations in this case is not known. Stotski and Seymour insist that Wycoff knew of the irregularities. "We have a total mess, and Wycoff took off for California," reports Patricia Grimm, a lawyer for Stotski.[16] The case ended in tragedy and murder in 1990. Only hours after winning custody of

Tessa, Reams was shot to death inside his ex-wife's apartment, where he had gone to pick up the child. Seymour was sentenced to eleven years in prison for killing her husband.[17]

The Reams case was not the first in which murder followed a surrogate birth. In 1981, Diane Downs became a surrogate for Dr. Richard Levin, a pioneer baby broker in Louisville, Kentucky, who was generally regarded as responsible for the nation's first surrogate birth. As part of his program, Levin requires psychological screening of his would-be surrogates. The psychological test done on Downs, which was conducted by Louisville psychologist Paul S. Mann, was not promising. The psychologist warned that "her [Downs's] long range good may not be achieved by participation in the surrogate project." Levin rejected the advice and suggested that Mann's analysis was "tainted" by his bias against surrogacy. He accepted Downs as a surrogate. Downs gave birth to and relinquished her surrogate child in 1982. Then, using $7,000 of insurance money from a "suspicious" fire in her mobile home, Downs began her own surrogacy program in Tempe, Arizona. In 1983, on a dirt road in Oregon—one year to the day after she gave up her child—Downs gunned down her own three children. One child was killed, the other two maimed. Downs is currently serving a life sentence in a New Jersey penitentiary.[18]

Death has stalked surrogacy in other ways. Robert Risner is a baby broker and freelance writer who works out of his home in Michigan. In October 1986, Risner flew down to Houston to meet with a prospective surrogate, Denise Mounce, to whom he had offered $10,000. The two signed a surrogacy contract and completed a psychological form in the dining room of a Houston airport hotel. Denise Mounce was twenty-three, single, childless, and impoverished. She lived in government-subsidized housing in Houston, Texas, and worked at minimum-wage jobs.

Denise had a developmental abnormality of the heart. She was subject to sudden, unpredictable episodes of rapid heartbeat. This defect was not detected in the physical examination given her in October 1986 as part of the surrogacy process. After five artificial inseminations, Denise became pregnant in March 1987. She was ill and tired during most of her pregnancy. During the seventh month of her pregnancy, she experienced symptoms of her heart condition and alerted her obstetrician's office. No medical care or treatment was administered. Risner was notified and referred her to a cardiologist, who in turn asked her for $250 so that he could fit her with a heart monitor. Though her broker had promised to be liable for medical expenses, Denise was not able to come up with the money and never returned to the doctor. She died of heart failure six weeks later, only a short time before her due date. With her died the five-pound, twelve-ounce life still in her womb.[19]

Denise's mother, Pat Mounce, a Virginia native, knew nothing of her daughter's financial crisis or contract pregnancy. She is currently suing the baby broker agency and the two physicians involved in her daughter's death. Pat Mounce has also become one of the nation's most articulate voices opposing contract childbearing. Risner, whose past is also riddled with fabrications and misstatements, has closed his surrogate "business" under a flood of lawsuits and state investigations. In May 1992, he was convicted on numerous counts of defrauding clients seeking to purchase children through his service.[20]

Scandal has been associated with virtually every surrogate broker's business. Maryland broker Harriet Blankfield suspended her business in March 1988 after seven years of baby brokering. Blankfield was one of the most visible baby sellers, repeatedly writing editorials and making radio and TV appearances in order to propagandize the benefits of surrogacy. Few knew, however, that her surrogacy program had been the subject of several complaints to the office of the Maryland attorney general. Included in those complaints were those of two client couples who claimed that they were double-billed by Blankfield. According to several women who were part of her program, Blankfield was abusive, insensitive, negligent, and dishonest. One mother in Blankfield's program, Cynthia Custer, canceled her contract with Blankfield because the broker refused to allow her any communication with the biological father who was to gain custody of her child. Custer did not wish to keep her child, but could not in good conscience give the child up unless she knew it was going into a good home. Blankfield, who had met Custer only once, refused all of her requests; in order to punish the recalcitrant surrogate, canceled her medical insurance for the surrogacy birth. Finally, after the intercession of other brokers from across the country who were frightened that Blankfield's misdeeds would further besmirch surrogacy, Custer was able to meet the father of her child. Only hours after its birth, Custer gave up her baby to the client couple in a roadside exchange, with consent papers signed on the hood of a car. Custer described her experience with Blankfield as "a nightmare," a description shared by several others in Blankfield's program.[21]

Blankfield is only one of thirteen surrogate brokers who have had to close over the last decade due to controversy, legislation, or litigation. In one case the Hagar Institute of Wichita, Kansas, declared bankruptcy and fled its location in the same month that a lawsuit was filed against it by a Colorado couple. The couple claimed that the surrogate mother provided by the institute had passed on a bacterial infection to the child they had bought. As a result of the infection, the baby became profoundly deaf.[22] In the same year that the Hagar Institute closed, two Michigan programs folded while their businesses were under investigation by adoption and postal officials.

The two most prominent surrogate brokers in the United States, William Handel and Noel Keane, have also not escaped controversy or the law. William Handel and his Center for Surrogate Parenting in Beverly Hills, California, created a public outcry by contracting American women to bear babies for numerous Japanese clients—clients who cross the Atlantic because surrogacy is illegal in Japan. Handel has also been sued by several dissatisfied clients and by at least one surrogate mother. Noel Keane, often termed the "big daddy" of surrogacy, has been responsible for more surrogate births than any other broker: over four hundred. Keane's original principal place of business was in Michigan. However, anti-surrogacy legislation passed in 1988 forced him to move to New York. Subsequently, in July 1992, New York also passed legislation prohibiting contract childbearing, forcing Keane to move again. Now he operates out of an office in northern California.

Keane's business has also been the target of numerous lawsuits. He was the broker in at least ten intensely contested cases, including the Baby M case. He has been sued for negligence and a variety of other misdeeds by a number of surrogate mothers, including Mary Beth Whitehead, Judy Stiver, Laurie Yates, and Patty Foster. One such litigant stated, "I don't want Noel Keane to be operating.

How many other women are going to suffer because of him?"[23] He has also been sued for malpractice by one client couple, after they discovered that they had spent $50,000 for a child who was not theirs.[24] Keane has the record for most children abandoned as a result of his business. According to the *Detroit News*, at least five children born in Michigan through Keane contracts have ended up in state-funded foster care.[25]

Keane is not remorseful or apologetic about the human tragedy and confusion his program has caused. He seems content to collect fees and let the courts settle the personal and legal chaos he creates. As one client noted, "He's just after the bucks, if you've got some green, you get in."[26] Keane's surrogacy operation yields an annual gross income of approximately $1,000,000.[27] Surrogate mothers have called Keane's program unprofessional and incompetent and stated that they are treated callously, as mere commodities in his "business deals."[28] Keane's apparent disregard for any basic ethical standards was in part responsible for the passage of anti-surrogacy legislation by the Michigan and New York legislatures, two of the nation's strongest laws against surrogacy.

The Baby Buyers

Many who come to baby brokers have chosen surrogacy as a last resort. These couples have struggled with years of infertility and often unsuccessfully attempted a variety of high-tech medical options to become pregnant. The publicized stories of these couples have led to the common assumption that virtually all the customers for surrogate children are infertile couples. Not so. The brokers themselves report that clients have included couples who, for reasons of health or employment pressure, simply do not wish to bear a child. Other clients have included unmarried couples, single men, and homosexual male and female couples.[29]

Remarkably, most brokers do not psychologically screen their clients to see if they are fit parents. Nor do they check their backgrounds for past history of child abuse or other criminal behavior.

According to the Office of Technology Assessment (OTA), none of the brokers they contacted required a home visit to ascertain the environment that contract children will be going into.[30] Only half of brokers require clients to undergo physical examinations, and just two-thirds require a test for sexually transmitted diseases. About the only real investigation consistently undertaken by brokers of clients is done to assure that the clients can pay.[31]

Unlike the women they hire, most clients are well-off and well-educated. The OTA reports that 64 percent of clients have incomes in excess of $50,000. Most brokers report that from half to 80 percent of their clients have had graduate school education. By contrast, the OTA found that most surrogate mothers earn just above the poverty line, and less than 4 percent of surrogate mothers are reported to have received graduate school education.[32] Over 40 percent of surrogates are unemployed, receiving financial assistance, or both.[33]

However, as demonstrated in several recent cases, money and education are no guarantees that baby buyers are stable. The tragic Reams-Seymour case described above is only one of several cases where clients have divorced soon after signing surrogate agreements. In one reported case, a divorcing client couple successfully pressured a contract mother to abort the contract child.[34] Equally disturbing is how many baby buyers act as if they were purchasing a mere product, showing an apparent total lack of normal human feeling toward the others involved in this tragic exchange of life. This troubling behavior made national headlines with the case of Patty Nowakowski. Nowakowski signed her surrogate contract in July of 1987. Noel Keane was the broker and a wealthy Detroit attorney the client. It was a typical Keane contract. It required Nowakowski and her husband Aaron "not [to] form or attempt to form a parent-child relationship with any child or children Patricia Nowakowski, Surrogate, may conceive, carry to term and give birth to. . . ."[35] As events turned out, the Nowakowski's inability to obey this clause of the contract avoided a heart-wrenching family tragedy.

In the fourth month of Patty's contract pregnancy, ultrasound revealed that she was carrying twins. Twins have become relatively common in surrogacy, as many brokers administer fertility drugs to surrogates prior to inseminating them, hoping to increase the odds of a quick conception. At first, the client couple seemed delighted.

Then, two weeks before the due date, the clients visited Patty's home and dropped a bombshell. They stated that they would only accept a girl child, not a boy child. The couple already had three boys and did not want another. They categorically stated that they refused to accept any responsibility for a boy—or two boys. Patty was heartbroken: "I went to my bedroom and started to cry uncontrollably. Hoping it would calm me down, I took a bath. But as I lay soaking my swollen belly, I could see the babies pushing under the skin of my belly and sobbed for their uncertain future."[36]

Patty and Aaron were in a bind. They already had three young children of their own and could not easily handle any additional children. In desperation Patty called Keane for help. He gave her none. The baby broker informed her that the only option was to let the couple have the girl, if one of the twins was a girl, and then have the boy (or boys) put up for adoption. In April 1988, Patty gave birth to a girl and a boy. Patty was uneasy and depressed in the hospital; "I wanted desperately to go home—but I knew that once I did, I'd never see my babies again." A couple of days later, time had run out.[37]

As they had indicated, the couple took only the girl and left her twin brother to be adopted. When picking up their baby, the clients seemed untouched by the fact that they were permanently separating sister from brother as well as daughter from mother. As recalled by Nowakowski, "Although they were elated and genuinely thankful for their daughter, they showed no remorse about leaving their son."[38] That same day the baby girl's twin brother was left with the adoption agency to be placed in a foster home. Days of agony passed for the Nowakowskis. Unlike the client couple, they could not forget the week-old infant boy, alone, about to be adopted by strangers. Finally, the Nowakowskis, despite their three children and financial worries, decided to take back Patty's son. The boy, who they named Arty Jay, soon became a loved member of the Nowakowski household. Arty's presence made Patty and Aaron feel even more strongly that it was shameful that their young son would grow up without his twin sister. Within a few weeks, the Nowakowskis took legal action against the client couple, asking for custody of their little girl. Facing a difficult and embarrassing court case, the couple relinquished the child. Six weeks after their birth, the twins were reunited. In August 1988, Aaron officially adopted Arty Jay and his sister, Alyssa.

Despite the happy ending to her story, Patty is unequivocal about surrogate motherhood. "I now firmly believe that surrogacy is not in the best interest of the children involved."[39] Nowakowski has testified in several state hearings on surrogacy legislation.

A Form of Bioslavery

Mothers like Patty who have had to give up their children only hours after birth as part of the surrogacy agreement have been psychologically and emotionally scarred. Surrogates have vividly described the overwhelming sense of loss they felt when confronted with losing their child. One mother has spoken of praying every night that she would not go into labor so that she and her baby could stay together.[40] Another surrogate stated: "Where there is no real baby, it is easy to be idealistic. . . . I started to grieve when I felt its movements. . . . [N]o amount of money can compensate."[41] Mary Beth Whitehead was handcuffed and taken into custody for the "crime" of wanting to keep her baby.[42]

The fact is that giving birth to a baby is not like producing a product. While some surrogate mothers are able to relinquish their "contract" children with few qualms, others cannot. They become emotionally attached to their children during gestation and birth. The now precious infant is no longer a legal abstraction in a contract. Many women cannot give up these children as if they were relinquishing a computer or refrigerator. For them maternal bonds often are far stronger than commercial contracts.

Remarkably, surrogate brokers operate on the basis that a mother can give up a baby immediately after birth as easily as a manufacturer turns over a product produced for a customer. Many contracts, like Keane's in the Nowakowski and Baby M cases, contain a provision requiring that surrogate mothers not form or attempt to form a maternal bond with their children.[43] This callous attempt to contract the destruction of the maternal bond is compounded by the brokers' repeated assertions that they have the capability to "screen" from their programs mothers who will feel bonds to their children.[44]

The emotional trauma experienced by surrogate mothers has been treated somewhat cynically by the surrogacy industry. Dr. Philip Parker screens surrogate applicants, counsels them, and does research on them. He thinks that emotionally disturbed women make "better" surrogates. Parker writes that "women with more neurotic or psychotic motivational factors or personalities" might experience surrogate trauma as "positive" or "helpful." He feels that surrogacy is a way for women to "master . . . guilt that they feel from past pregnancies that ended in abortion or adoption." Approximately 9 percent of the surrogates he examined felt they were atoning for a previous child relinquished for adoption; 26 percent felt that they were making up for an abortion.[45] As one twenty-three-year-old potential surrogate who had previously aborted a child explained, "I killed a baby. Now I could make up for it by giving one to a needy family."[46]

The psychological damage done by surrogacy could be long-term. In a comprehensive survey of women who had given up their children for adoption, the majority of the women felt an overwhelming sense of loss for periods extending up to thirty years.[47] Surrogacy contracts, if enforced, will create a new group of grieving women. For unlike mothers in the adoption circumstance, who have willingly given up unwanted children after their birth, many surrogate mothers are forced to give up children that they want, based on an uninformed decision made before they conceived the child.

The physical and emotional exploitation of women resulting from surrogacy has had a disproportionate impact on economically disenfranchised women. These women, who are overwhelmed by immediate or impending economic hardships, are far more likely to enter into agreements requiring them to relinquish their parental rights. They do not realize, or simply disregard, the risks of physical and emotional harm that may result from carrying out the arrangement. Several of the surrogate mothers who have made their cases public have openly discussed the economic pressure to become surrogates.[48] Brokers take full advantage of this economic coercion. Dr. Howard Adelman of Surrogate Mothering, Ltd., who routinely screens surrogates, notes that women in financial need are the "safest" surrogate applicants. Their need for money makes them less likely to change their mind after signing a surrogate arrangement.[49]

Given their economic need, the compensation offered for the performance of a surrogacy contract, generally around $10,000, is the deciding factor for many potential surrogate candidates. A recent study illustrates this point. It indicates that, while many women enter into surrogacy contracts with the altruistic intent of providing an infertile couple with a child, the vast majority of women interviewed would not participate in a surrogacy arrangement unless they received a fee. Surrogacy is baby-selling, pure and simple.[50] The attorney general of Michigan puts it plainly:

> **The money plaintiffs seek to pay the "surrogate" mother is intended as an inducement for her to conceive a child she would not normally want to conceive, carry for nine months a child she would not normally want to carry, give birth to a child she would not normally want to give birth to and then, because of the monetary reward, relinquish her parental rights to a child that she bore.[51]**

Apparently, even the current economic incentive for surrogates is higher than the market might sustain. John Stehura, president of the Bionetics Foundation, Inc., which helps arrange surrogate transactions, has predicted that corporations such as his will be able to recruit poor women both in the West and Third World countries for a fraction of the current rate.[52]

In the face of the exploitation and extraordinary ethical problems involved in surrogate motherhood, most nations in the Western world have banned the practice. The health minister of France has declared surrogacy "slavery over women." Germany forbade broker Noel Keane from operating an office within its borders. Other countries, including Australia, Israel, Norway, Spain, Switzerland, and the United Kingdom, have banned surrogacy. Public policy organizations in Austria, Canada, Italy, the Netherlands, New Zealand, and Sweden, as well as prestigious international organizations such as the Council of Europe and the World Medical Association, have also rejected commercial surrogacy.[53]

In the United States, it has proven far more difficult to ban contract childbearing. The United States is a country deeply committed to the ideology of the market. Many U.S. legal scholars and economists defend surrogacy on the grounds that if a market system is to survive, contracts must be sacrosanct, more so, apparently, than even motherhood. They also argue that an open market in babies would enhance public good by more equally distributing babies from those who have them (often the poor) to those who do not and can afford to buy them. Free market advocate and now appellate court judge Richard Posner has argued that the sale of children actually increases their welfare: "[The] willingness to pay money for a baby would seem on the whole a reassuring factor from the standpoint of child welfare. Few people buy a car or a television set in order to smash it. In general, the more costly a purchase, the more care the purchaser will lavish on it."[54] Posner neglects to note in his analysis that when most consumers buy a durable good, such as a car or television, they "junk it" when its useful economic life is over or when a newer model is more attractive to them—a frightening precedent if applied to children. Moreover, many commentators flatly reject the idea that wombs can be rented or children bought under the laws of supply and demand. Katha Pollitt writes, "Goods can be distributed according to ability to pay or need. People can't. It's really that simple."[55]

The struggle between the traditional concepts of the intrinsic dignity of motherhood and laissez-faire market ideology has not been legislatively resolved. Though numerous anti-surrogacy laws have been introduced, the U.S. Congress has failed to pass any prohibitions on commercialized childbearing. However, there is a growing trend in state legislation to prohibit payment and contracts for childbearing.[56] Eighteen states now restrict surrogate parenting in whole or in part.[57] Moreover, recent court cases show a strong trend toward voiding surrogacy contracts.[58]

Recent legislative and judicial decisions indicate that surrogacy in the United States may be on the wane. However, in one area of commercialized childbearing, nongenetic surrogate motherhood, the current legal situation is not moving toward prohibition, but rather toward full legal acceptance of maternity for hire. In fact, if current legal opinions hold up, we may see a legal basis set for turning women into commercial breeders and a historic and frightening shift in the legal definition of motherhood—a definition unique in the history of western law.

The Human Oven

Many surrogacy contracts involve women being implanted with the fertilized egg—the embryo—of couples. They are paid up to $10,000 to bring these embryos to term.[59] To distinguish them from the majority of surrogate mothers who have been artificially inseminated, these women are called nongenetic, or gestational, surrogates. The first contracted nongenetic surrogacy was negotiated by Noel Keane in 1986. Since Keane's action, numerous other brokers have begun contracting nongenetic surrogate mothers. In that the gestational surrogate has by definition no genetic link to the child, brokers and clients are beginning to solicit women from minority groups who could become gestational surrogates at low cost. Additionally, brokers favor minority and poor women as nongenetic surrogates because they feel it is unlikely that such women will have the resources to fight for their children. They also believe that the U.S. judicial system, at times racially biased, is less likely to award minority surrogates custody over white well-to-do couples.

Are these women, in whose bodies the embryo of others is implanted and brought to term, legally birth mothers? Despite the lack of a genetic tie, does their intimate emotional and physical attachment to the child make them the same as any other mother? Should their providing the developing fetus with blood, nutrients, and other key body substances entitle them to some basic parental right? According to two California courts, the answer is no. These courts have become the first in the history of western jurisprudence to hold that a birth mother has no parental rights—that she cannot even be called a mother. The case that these two courts ruled on, the Anna Johnson case, may become one of the most important legal decisions of the century.

Even by human body shop standards, the contract signed by Mark and Crispina Calvert and Anna Johnson on January 15, 1990, was unusual. The contract provided that an "embryo" created by fertilizing the egg of Crispina with the sperm of Mark would be implanted in Anna. The agreement also stipulated that Anna, who is African-American, give up to the Calverts the child who was to result from this process. For her services Anna would be paid $10,000 in a series of installments, the last $5,000 to be after the child's birth and relinquishment of the infant to the Calverts. The contract had come about because a tumor in Crispina's uterus had forced her to have a hysterectomy in 1984. The operation, however, left her the ability to produce eggs, and the couple eventually considered nongenetic surrogacy. In 1989, Anna heard of the Calvert's plight and offered to help.

The fertilized embryo was implanted into Anna on January 19, 1990. Anna had been taking a variety of hormones in order that her body would not reject the fetus. Even with medication, the odds of embryo transfer working are not very good—generally less than 15 percent. But about a month later, an ultrasound test revealed that Anna was pregnant. After the pregnancy started, relations between the Calverts and Anna deteriorated quickly. In August, Mark and Crispina filed a lawsuit seeking a declaration from the court that they were the legal parents of the unborn child. Anna filed her own action, seeking a declaration that she was the mother of the child. The child, a baby boy, was born in September 1990. For a few days, both sides fought for temporary custody of the child, whom Anna called Matthew and the Calverts called Christopher. Finally, in order to reduce stress on the child, Anna allowed the Calverts custody on a temporary basis as a "foster" home, and she retained visitation rights. In mid-October, the case went to trial. Several days of testimony followed, including testimony on intrauterine bonding—the bonding between the mother and the child she is carrying, regardless of her genetic link to the child.[60]

On October 22, 1990, in front of a packed courthouse in Orange County, California, Superior Court Judge Richard N. Parslow issued a rambling, folksy oral opinion. Hidden in his asides ("I see, as they say in the business, we have a packed house today") and in his attempts at humor ("Artificial insemination. . . . The last I heard, they were doing some of those with turkey basters. . . ."), Parslow made history.[61] The judge held that Anna Johnson was not the mother of the baby boy to whom she had given birth. He decided, citing no law or cases for support, that the age-old definition of mother had become superseded by technology; that in nongenetic surrogacy the birth mother was no longer a mother. Rather, she was part of a new category of woman essentially invented by the judge: "Anna Johnson is the gestational carrier of the child, a host in a sense."[62] He declared Crispina Calvert the natural mother of the baby boy because of her genetic link to the child. Parslow stripped Anna of all parental rights and revoked her visitation.

Many commentators were deeply disturbed by Parslow's opinion. Ethicist James Nelson of the Hastings Center noted that Parslow's opinion "offers an image of women as interchangeable fetal containers, and that when it comes down to it, any womb will do."[63] Psychiatrist and surrogacy expert Dr. Michelle Harrison found the judge's view that Anna (the birth mother) was not the legal mother of her child incomprehensible:

> Home is where the womb is. For the slowly evolving fetus, home is the place filled with the mother's warmth and cushioned by her fluids. . . . In the last months, the fetus hears her singing, her talking, her crying. At birth the newborn shows preference for her voice above all others. . . . The donor of the egg did not experience the nine month long intimate, dynamic and life giving process that went on for every second of gestation. In the fullest biologic sense, the donor of the egg has not mothered that baby.[64]

Paradoxically, many proponents of advanced reproductive technology were also concerned about the court decision. Fertility researchers had just announced that they had achieved pregnancy in women past menopause by getting these women implanted with donated embryos. Parslow's decision would make the ovum donors, not the postmenopausal women, the real mothers of those children.[65]

The lower court decision was quickly appealed by Anna's attorneys, Richard Gilbert and Diane Marlowe. Expectations were high that the ruling would be reversed. However, on October 8, 1991, the California Court of Appeals upheld Parslow's holding and also denied Anna any parental rights. Disappointed once again, Anna's attorneys then appealed the decision to the California Supreme Court. On May 20, 1993, California's top court came down with a sweeping endorsement of gestational surrogate motherhood. The court firmly backed using commercial law concepts to force women to obey rent-a-womb contracts. Anna, according to the court, had no rights as a mother. The court could see no special status for pregnancy and childbirth as compared to other services for sale. The court saw no legal distinction between women being paid for childbearing or those contracted to perform "lower-paid or otherwise undesirable employment."

In an impassioned dissent Justice Joyce Kennard, the lone woman on the court, chided the majority approach that "entirely devalues the claims of motherhood by a gestational mother such as Anna." "[T]he gestational mother's biological contribution of carrying a child for nine months and giving birth is an assumption of parental responsibility," Kennard writes. "A pregnant woman's commitment to the unborn child she carries is not just physical; it is psychological and emotional as well." Justice Kennard was also deeply disturbed by the majority's reliance on property and commercial law to decide the future of Anna's child. "[In deciding Anna's rights as a mother] the majority has articulated a rationale . . . grounded in principles of tort, intellectual property and commercial contract law. But as I have pointed out, we are not deciding a case involving . . . the delivery of goods under a commercial contract; we are deciding the fate of a child."[66]

After the defeat in the California Supreme Court, Anna's case was appealed to the U.S. Supreme Court on constitutional grounds. But on October 4, 1993, the Court refused to review the decision. Anna's legal struggle was over.[67] She became the first birth mother in the history of Western law to be denied the legal standing of "mother." Richard Gilbert found the California court's ruling deeply disturbing: "We've lost our way as humans when we begin to treat our birth mothers as objects or commodities in the marketplace. Clearly birth mothers are parents and I fear for a legal system which 'plays God' in denying them this natural and legal right."[68]

———

The California courts' unprecedented separation of the legal definition of motherhood from the act of giving birth allows motherhood to be sold as a commercial service. It opens a brave new world where women will have sold various functions of their bodies to produce a commodity, in effect creating a new commercialized method of childbearing. As technology develops, the "surrogate" becomes a kind of reproductive technology laboratory, a serviceable "maternal environment" for the purposes of bearing the "customer" couple's child. In short, she has been dehumanized and reduced to a production machine, a "factory" producing the human body shop's most precious product.

9

This whole business seems to treat human reproduction as something analogous to a car. Do we want it to be bright or blue? How do we make it sleeker? It turns reproduction into a technological procedure, more and more divorced from human intimacy and relationships.

Jean Bethke Elshtain[1]

The Perfect Baby

AT BIRTH, CHLOE O'Brien seemed to be no different than any other healthy six-pound baby girl, but appearances were deceiving. Born in March 1992 at Hammersmith Hospital in London, fourteen years after Louise Brown, the world's first "test-tube baby," Chloe was the pioneer product of the remarkable marriage of in vitro fertilization and the cutting-edge science of genetic diagnosis. Chloe was the first baby to be genetically screened as an embryo for a genetic defect, cystic fibrosis (CF), before being implanted into her mother's womb.[2]

Chloe's beginnings were quite complex. The medical team at Hammersmith Hospital (which included Mark R. Hughes, director of the Prenatal Genetics Center at Baylor College of Medicine in Houston, Texas) removed several eggs from Michelle O'Brien's ovaries and placed them in a sterile dish filled with protective fluids. The ova were then fertilized with the sperm of her husband, Paul. Cell division began in the fertilized eggs. Each embryo split to form two cells, then four, then eight. At the eight-cell stage, researchers performed a procedure on each embryo called extrusion. Using a microscopic hollow needle, they punctured the "shell" surrounding each embryo and plucked out one of the embryo's cells. The removed cell provided the basis for the CF test.

A single cell is not enough material on which to conduct genetic screening. So researchers "copied" the cell thousands of times, using a technology known as PCR (polymerase chain reaction). With this amount of test material, doctors were able to tell which of the embryos had the CF genes and which did not. Embryos with both CF genes were discarded. Several embryos with normal genes or with only one CF gene were kept for potential implantation into Michelle O'Brien. Ultimately, two embryos were implanted, one with two normal genes, one with one normal gene and one CF gene. Only one of the embryos developed, and several months later Chloe was born.[3]

Chloe's parents had chosen to undergo in vitro fertilization and then the genetic screening of the resulting embryos because their son Martin, who was born in 1988, has CF. Both Michelle and Paul carry the CF genes, and their children have a one in four chance of being born with the disease. A child must inherit a defective gene from both parents in order to develop the genetic disorder. They wanted to make sure that their next child did not inherit both genes.[4]

The Hammersmith team is only one of several teams around the world that are working on preimplantation genetic screening of early embryos. Researchers at the Howard and Georgeanna Jones Institute for Reproductive Medicine in Norfolk, Virginia, have similar ambitious plans for the genetic screening of preimplantation embryos. The Institute's forty-member research team, led by pregnancy specialist Gary D. Hodgen, is planning to begin a U.S. embryo screening program by 1993.[5] The Virginia researchers have targeted Tay-Sachs disease, an inherited fatal disorder found primarily among Jews of Eastern European ancestry, to be the first disease tested. Another Virginia clinic, the Genetics and IVF Institute, has also announced that it will begin genetic testing of IVF embryos. Its program will target a number of disorders, including cystic fibrosis and sickle-cell anemia. "We are actually ready to go," says Susan Black, the institute's clinical director for preimplantation genetics.[6] Additionally,

Dr. Yuri Verlinsky and his partner, Dr. Charles Strom, have made Chicago's Illinois Masonic Medical Center an international center for preimplantation genetics. In 1990, the Illinois Center sponsored the First International Conference on Pre-Implantation Genetics, attended by more than 250 researchers from around the world.[7]

The increase of interest in and use of preimplantation genetic screening indicates that a highly profitable new reprotech industry is on the horizon. Undoubtedly, the future will see more and more customer couples such as Michelle and Paul O'Brien using in vitro fertilization and genetic screening as a way to choose which genetic traits they want their children to have. Chloe will be only the first of a new wave of genetically tested "test-tube" babies.

Genetic testing of fetuses did not, of course, begin with Chloe or with preimplantation genetic screening of embryos. Prenatal diagnosis started in the 1960s with the development of amniocentesis. In this procedure the doctor inserts a long needle into the amniotic cavity of the womb and extracts some of the amniotic fluid, which contains fetal cells. Through microscopic examination of these cells, technicians can discover up to two hundred genetic disorders in the developing fetus. The procedure is not risk-free. Estimates on risk to mother or fetus vary from about 0.5 percent to 1.5 percent. The procedure has another major drawback: attaining genetic information on the fetus through amniocentesis requires that the fetus be between sixteen and twenty weeks of age. Should the fetus have a serious genetic defect, this amount of time can often push the abortion decision to late in the second trimester, causing ethical and medical dilemmas for parents. In 1970, only two hundred amniocentesis fetal screenings were attempted. Now, a little over twenty years later, more than 300,000 women undergo amniocentesis each year.[8]

A new, highly touted genetic diagnostic technique called chorionic villus sampling (CVS) brings results more quickly than amniocentesis—as early as nine weeks after conception. The procedure, which is guided by ultrasound imaging, involves going through the vagina to snip material from the developing fetal sac. Then, just as in amniocentesis, this material can be examined for a wide variety of genetic and chromosomal abnormalities. However, persistent questions about the safety and accuracy of this procedure for fetus and mother have restricted its general use.[9]

Unfortunately, cures for the genetic diseases being screened for in amniocentesis and CVS simply do not exist. Though half of currently known genetic diseases are fatal and the vast majority are seriously disabling, only 15 percent can be effectively treated. Until cures are found, parents confronted with afflicted children have only two choices—abort the abnormal fetus (often late in the pregnancy), or live with the disease. Making this choice is a confusing and frustrating process. While genetic counselors often spend up to two hours with each patient, there are only approximately one thousand such counselors in the United States. The vast number of patients will rely on the advice of the family doctor, who often will have no more than a few minutes to spend with them. This is potentially tragic, since with fetal screening the consultation with a counselor or doctor often will determine a couple's choice on whether to terminate a pregnancy.

In the worst case, couples may opt for abortion because they misunderstand the odds of their child being born with a particular condition. One study of middle-class pregnant women, for example, found that one-quarter of them interpreted "1-in-1,000 chance" to mean a 10 percent or greater chance.[10] During the 1970s, there was widespread confusion among African-Americans, and among policymakers, over the meaning of genetic tests for sickle-cell anemia. Many couples sought abortions, thinking that they would pass on sickle-cell anemia when they carried only one copy of the gene, not the two required to get the disease. The testimony of Johns Hopkins University professor Neil A. Holzman, who has studied risk communication by doctors, is not reassuring. "I'm not sure," says Holzman, "we are confident we can communicate risk accurately to people."[11] Moreover, many couples do not have sufficient knowledge about the diseases being tested. A couple's decision to abort a fetus with Down's syndrome, CF, Huntington's disease, or other long-range illnesses may depend on how the genetic counselors or doctors describe the disease—whether they describe it as a nightmare, or as a chronic or future disorder to which the families can adjust.

As compared with amniocentesis or CVS, preimplantation genetic screening of embryos provides couples with a revolutionary option. Parents can find out about genetic disorders before the embryo is implanted in the womb. Rather than being aborted several weeks or months into the pregnancy, abnormal embryos are simply discarded. As the techniques of preimplantation screening evolve, they may represent the ultimate tool in the genetic diagnosis of the unborn.

However, the genetic screening of preimplantation embryos also brings up troubling concerns about the legal and moral status of embryos. Is there any limit to how many "inferior embryos" we can discard, or how often? And for what reasons is it permissible to destroy embryos? Do they have to be afflicted with serious genetic diseases, or can they be discarded for reasons having little to do with disease, such as the fact that the embryos may be of the "wrong" sex, or genetically predisposed to obesity or low I.Q.? These questions have not been and are not being answered. As with so many repro-tech advances, the new embryo screening technology is totally unregulated. The absence of legal limits on embryo screening leaves the door open to the creation of a new body shop business designed to provide couples with the "perfect baby."

Eugenics and Other Evils

As we enter the world of preimplantation genetic screening of embryos, there is increasing evidence that genetic testing is already being done on the unborn in areas that have little or no relation to serious disease. Currently, there is growing use and acceptance of amniocentesis and CVS for a pernicious eugenic purpose; sex selection of the unborn. Sex selection abortion, once thought to be the province of a limited group of ethnic populations, is being used and accepted at a greater rate than many would have thought possible only a short time ago. National surveys taken in 1973 and 1988 indicate that the percentage of geneticists who approve of prenatal diagnosis for sex selection rose from 1 percent in 1973 to nearly 20 percent in 1988.[12] Geneticists say that this significant change in attitude is caused by the greater availability of prenatal diagnostic technology and by increasing numbers of patients asking for the tests.

A recent international study underscores the growing acceptance of sex selection abortion. Among its other findings, the study describes the results of a poll taken among doctors when presented the following scenario:

> A couple with four healthy daughters desires a son. They request prenatal diagnosis solely to learn the fetus's sex. . . . They tell the doctor that if the fetus is female, they will abort it. Further, they say that if the doctor will not grant their request for prenatal diagnosis, they will have an abortion rather than risk having a fifth girl.

When presented with this case, 62 percent of 295 U.S. doctors said that they would either perform prenatal diagnosis (34 percent) or would refer them to someone who would perform it (28 percent). Geneticists in some other Western nations are also showing willingness to perform prenatal sex selection. When presented with the scenario described above, sizable percentages of these doctors said they would either perform prenatal diagnosis for the couple or refer them (United Kingdom, 24 percent; Greece, 29 percent; Brazil, 30 percent; Israel, 33 percent; Sweden, 38 percent; Canada, 47 percent; and Hungary, 60 percent).[13]

Internationally, India is the worst offender in performing sex selection abortions. In a country that has no old-age pensions, sons are prized as a financial support for families, while daughters can be a financial liability due to complex dowry-like marriage arrangements among many Indians. This social and economic preference for males has been disastrous for the female unborn. Government reports estimate that between 1978 and 1982, 7,999 out of every 8,000 abortions in the Bombay area involved female fetuses. A survey of fifty obstetricians in private practice in Bombay indicated that forty-two were performing amniocentesis for sex selection. India has begun responding to this crisis in eugenic abortions. In 1988, the Indian state that includes Bombay passed a law criminalizing the disclosure of a fetus's gender. Other Indian states are considering similar legislation.[14]

There are no figures on how many sex selection genetic diagnoses and abortions are performed each year in the United States. Clinics and hospitals are not required to keep records on the reasons for an abortion. But most researchers and doctors in the field feel that the practice is becoming far more prevalent. Numerous geneticists and clinics have reported that they regularly receive requests for prenatal diagnosis for sex selection. In one year, Massachusetts General Hospital in Boston received more than one hundred inquiries for chorionic villus sampling procedures just to determine sex. And the requests are not merely from Indians and other ethnic groups. "I've found a high incidence of sex selection coming from doctor's families in the last two years," reports Dr. Lawrence D. Platt, a geneticist at the University of Southern California. He notes that doctors' requests for the test are "much higher than ethnic requests. Once there is public awareness about the technology, other people will use the procedure as well."[15]

Even preimplantation genetic screening of embryos started with sex selection. Over two years before the birth of Chloe, Dr. Allan Handyside of London's Hammersmith Hospital used IVF and genetic screening to test embryos for sex identity. Reportedly, Handyside used sex selection because the genetic defects being looked for in the embryos only affected males.[16] However, while Hammersmith's use of sex selection was designed to prevent disease, the techniques could just as easily be used for simple sex selection. In the future, increasing numbers of couples may undergo IVF and genetic screening of the resulting embryos in order to be implanted only with the embryo that is the sex they desire.

The growing acceptance of sex selection sets a dangerous eugenic precedent for attempts to abort fetuses based on "undesirable" genetic characteristics such as low I.Q., short stature, or poor eyesight. Undoubtedly, in the future geneticists will be asked to screen for a number of undesirable traits that parents feel will affect the quality of their child's life and their own. If parents will screen babies for one nonmedical condition, that is, gender, there is no reason to assume they will not screen them for others. Results from a recent poll taken among New England couples graphically demonstrate the coming danger of a new eugenics. Among the couples polled, 1 percent would abort a fetus on the basis of sex, 6 percent would abort a child likely to get Alzheimer's disease in old age, and an incredible 11 percent would abort a child predisposed to obesity.[17]

Our knowledge of genetics and the ability to screen for a wide variety of genetic traits both related and unrelated to diseases is increasing rapidly. Researchers have already identified hundreds of genetic "defects"; the U.S. government is now sponsoring a $3 billion program called the Human Genome Project, which is designed to decipher all of the over 100,000 genes in the human body. U.S. researchers are not limiting their inquiries to genes related to serious disease. Recently, the National Institutes of Health (NIH) announced a grant of $600,000 to track the genes for I.Q. Psychologist Richard Plomin, who is heading the three-year project, hopes it will help better identify "the really smart kids."[18]

By the next century, many scientists expect that the information gained through the Human Genome Project will enable doctors to screen fetuses and test-tube embryos for an extraordinary variety of physical and behavior traits. Some predict that gene combinations for even the most trivial of traits could be found and screened for. One scientist noted with evident concern that "we should be able to locate which combinations affect kinky hair, olive skin and pointy teeth."[19] For the first time in history, parents could be deciding, not wondering, what kind of children they will bear and discarding those seen as imperfect or defective.

The genetic screening and testing of embryos for a variety of genetic traits could be a multibillion-dollar human body shop enterprise. Even with the current limited number of tests for genetic diseases, there is tremendous pressure on doctors and federal agencies to begin massive screening programs that funnel billions of dollars into new biotech screening companies. Referring to attempts by the U.S. government to begin a nationwide cystic fibrosis genetic screening program, one commentator noted, "This is potentially a billion dollar industry. The commercial molecular biology companies are pushing it hard. The pressure to test is very powerful."[20]

Under pressure from the genetic screening and medical industries, couples in the future may increasingly pick in vitro fertilization rather than natural childbirth. In vitro fertilization would allow doctors to have several embryos on which they could conduct "embryo biopsies" to determine which of the embryos had the most desirable genetic traits. Doctors would then reimplant the desired embryo or embryos in the mother or a paid surrogate mother. Parents could literally pick and choose which embryos have the characteristics that match their desires and discard the others. The women need not suffer through abortion, as with amniocentesis or CVS, and genetic trait diagnosis and selection could be refined almost infinitely. Parents may buy into this unnatural production method based on the hope of creating the "perfect" child. Of course, as demonstrated by the large percentage of parents who say they would abort a child predisposed to obesity, concepts of what is perfect and what is abnormal or bad are often merely reflections of the cultural stereotypes and prejudices of ourselves and our society. As noted by legal scholar George Annas, "The whole definition of normal could well be changed, the issue becoming not the ability of the child to be happy but rather our ability to be happy with the child."[21]

The human body shop implications of prenatal genetic screening have not been lost on practitioners. "I see people occasionally in my clinic who have a sort of new car mentality. [The baby's] got to be perfect, and if it isn't you take it back to the lot and get a new one," states Dr. Francis Collins, a geneticist who helped isolate and identify the gene responsible for cystic fibrosis.[22] "We do have in our society a premium baby mentality," says Mary Mahowald, a professor in the Department of Obstetrics and Gynecology at the University of Chicago. She continues,

> It is eugenics. We don't give it that name, but we foster the concept nevertheless. It has intensified over the last decade because of the two child family, the availability of abortion and the techniques we have for pre-natal, even pre-pregnancy, diagnosis. All those together contribute to the notion that people not only ought to be able to determine when to have children and how many to have, but also just what kind of children to have. . . .[23]

Probably all prospective mothers and fathers enter pregnancy and childbirth with fears that their baby will not be "all right." With the possible exception of death, humans may never feel so at the mercy of nature as when the mystery of birth unfolds. Attempting to control this process and avoid the tragedy of a badly deformed or ill child is both understandable and in many ways laudable. Now, however, along with the altruistic attempt to control fatal or serious hereditary disorders, we are confronting a new eugenics. The genetic screening of embryos and fetuses for nondisease traits is pushing the practice of human eugenics into a new and frightening arena. We no longer have political or racial eugenics as practiced earlier in this century, most notably in Nazi Germany, but instead have begun a "commercial" eugenics being peddled by clinics, researchers, and biotechnology companies for profit. We no longer are preventing the marriage of "undesirables," sterilizing the "unfit," or exterminating races viewed as inferior. But we are creating a new market in genetic trait selection of children; a business peddling the prevention of birth of those that do not meet the expectations of parents; an industry that would destroy the unborn if they do not fit into "the perfect baby" mold, be they female, predisposed to obesity, or potential victims of disease decades into their lives. We have not developed any legal or policy mechanism to prevent this terrible prospect, to limit the use of our diagnostic techniques so that they are only used for the detection of serious illness. Large vested interests are pushing against any such limits and for the wholesale extension of the human body shop mentality into the birth process. The new commercial eugenics of birth is every bit as frightening as the political eugenics of the past, only it will be far more difficult to identify and control.

To Be or Not to Be

One of the greatest fears about prenatal genetic diagnosis is that it will heighten personal and public intolerance of disabilities and disabled children. With the increasing ability to diagnose genetic diseases and disorders how will society feel about those who are born with such disabilities? Is a disabled life worth living, or would it be better if certain lives were not lived at all? If the answer is that certain lives are not worth living or are of diminished worth, should children born with genetic defects, and their parents, be able to sue doctors for failure to use genetic screening to prevent the birth of such children? These questions have puzzled and confounded the legal world over the last several years. As of 1990, over three hundred cases had been filed in dozens of states by parents or children claiming "wrongful life" or "wrongful birth."[24]

Among the first of these ground-breaking cases was *Berman v. Allan*, which went before the Supreme Court of New Jersey in 1979.[25] On September 11, 1975, Paul and Shirley Berman filed a malpractice suit against two New Jersey doctors, Ronald Allan and Michael Attardi. In their suit the Bermans alleged two different causes of action against the doctors. The facts behind the claims were straightforward. From February 19 until November 3, 1974, the pregnant Mrs. Berman, age thirty-eight, was under the care of Drs. Allan and Attardi. At no time during her pregnancy did her doctors recommend or conduct amniocentesis on the fetus. On November 3, the Bermans' daughter Sharon was born. She was afflicted with Down's syndrome.[26]

The Bermans asserted that had they known about Sharon's condition, they would have aborted her prior to birth. They alleged that given Mrs. Berman's age, the doctors should have informed her about the risks of having a Down's syndrome baby and should have made the test available to her. Because of the doctor's alleged negligence in not informing Mrs. Berman, the Bermans sought compensation for the pain and suffering that Sharon would endure during her "wrongful life." They also requested damages in their own right for "the emotional anguish" they had experienced and would continue to experience on account of Sharon's birth defect. They named this second claim "wrongful birth."[27]

The court denied the "wrongful life" claim filed on behalf of Sharon. At the outset the court noted that it was not able to affix damages for such a "metaphysical" claim. As stated by the court, "In the case of a claim predicated upon wrongful life, such a computation would require the trier of fact to measure the question of the difference in value between life in an impaired condition and 'the utter void of nonexistence.'"[28] The court continued, "Ultimately, the infant's complaint is that she would be better off not to have been born. Man, who knows nothing of death or nothingness, cannot possibly know whether that is so."[29] Additionally, the court could not bring itself to call birth, including birth of a disabled child, a "damage" to that child. As noted by Justice Pashman in his decision, "One of the most deeply held beliefs of our society is that life— whether experienced with or without a major physical handicap—is more precious than non-life."[30]

The New Jersey court's dismissal of "wrongful life" cases has become the norm for the numerous such cases filed across the country. States have also acted against these claims. By 1991, six states had passed laws limiting or forbidding recovery of compensation for cases of wrongful life.[31] But the verdict is not unanimous. In the 1980s, three courts recognized the child's ability to bring a wrongful life suit.[32]

The New Jersey Supreme Court was more forthcoming toward the Berman's claim for "wrongful birth." The court held that doctors owed parents a duty to provide them with genetic testing and screening that would aid them in making the abortion decision. Justice Pashman reasoned that:

> . . . a physician whose negligence has deprived a mother of this opportunity [abortion] should be required to make amends for the damage which he has proximately caused. Any other ruling would in effect immunize from liability those in the medical field providing inadequate guidance to persons who would choose to exercise their constitutional right to abort fetuses which, if born, would suffer from genetic defects.[33]

The court did not award medical costs to the Bermans for raising Sharon, reasoning that this would not be appropriate given the "love and joy" that the child offered. However, the court did feel it appropriate that the Bermans be awarded "emotional damages" for the "mental and emotional anguish" caused by their realization that they had given birth to a child afflicted with Down's syndrome.[34]

Since the *Berman* decision, several courts across the country have allowed for recovery against doctors who failed to administer genetic testing of fetuses when such testing seemed called for. This includes instances where genetic counseling was not given to Jews susceptible to Tay-Sachs disease, families with a prior history of cystic fibrosis, or African-Americans prone to sickle-cell anemia. Courts have also held against doctors and clinics who were guilty of errors in genetic testing that resulted in not diagnosing genetic defects.[35] Only one state, Missouri, has banned lawsuits based on wrongful birth.[36]

As the case law remains in flux, ethicists also are at odds over the morality of wrongful-life and wrongful-birth cases. "I can certainly conceive of situations where it would be better not to have attained consciousness," says Leroy Walters, director of the Center for Bioethics at the Kennedy Institute of Ethics at Georgetown University. "In that sense, I think there is some ethical foundation for bringing wrongful-life suits."[37] "Sometimes, as with Tay-Sachs, there is a fate worse than death," states Alex Capron, professor of law at the University of Southern California. "And then the question becomes, 'What is this child's suffering worth?'"[38] Douglas Johnson of the National Right to Life Committee strongly disagrees. He supports abolishing the new causes of action:

> They [wrongful-life and wrongful-birth cases] infringe on the freedom of conscience of physicians. If a physician is seeing a pregnant mother, he is responsible for two patients: the mother and the fetus. If the mother is over 35 and the doctor doesn't tell her about amniocentesis and Down's syndrome, he runs the risk of being sued. He has a legal gun to his head, coercing him to offer a procedure with no therapeutic value. He is forced to participate in eugenic medicine, that which simply tries to improve the human gene pool, something that fell out of favor after the Third Reich. Killing a patient is not a form of therapy.[39]

In the new age of preimplantation genetic screening of embryos, it is unclear how far a physician's duties go. Is the physician obligated to inform patients of available tests to avoid genetically abnormal or defective fetuses? Is the physician liable for children born predisposed to obesity, poor eyesight, or low I.Q., assuming the availability of such tests in the near future? It is clear that as the frontiers of genetic diagnosis and embryo manipulation become routine medical practice, these legal and ethical questions will be increasingly asked of courts in the United States and around the world. And if the courts legally affirm concepts of wrongful life or wrongful birth, they impose on the medical profession an obligation, perhaps a duty, either to use prenatal genetic screening, or at a minimum to advise parents of the existence and functions of such tests. Widespread genetic screening, already propelled by the profit motive, will also be boosted by physicians' fears of liability lawsuits. Moreover, as the current reproductive technologies offer couples hopes of "perfect" babies through the "miracles" of medicine, expectations will continue to be raised. Patients will feel even more wronged by the birth and the lives of children considered to be diseased, inferior, or unfit.

The new trends toward creating a perfect baby may represent the ultimate in the baby factory, but the genetic screening of embryos and fetuses also signals a new body shop revolution spawned by the rapidly expanding biotechnology industry. Sophisticated techniques in genetic diagnosis and engineering are barely two decades old, yet they have already created a boom market in a new set of biological commodities: the over 100,000 genes of the human body. The next section will describe the ongoing genetic engineering revolution and will assess what the startling advances in the field could mean for the commodification and alteration of humankind.

PART THREE

The Gene
Business

10

We need wisdom most when we believe in it least.
Hans Jonas[1]

Designing Genes

THROUGHOUT THE INDUSTRIAL age, we have radically altered our society and environment by collecting and transforming massive amounts of inanimate material. Numerous metals, minerals, and fossil fuels are mined, filtered, dug, and pumped from the earth. They are then burned, forged, soldered, melted, restructured, and recombined to create the machines, structures, and artifacts of the modern world.

We are now adding living human material to the inanimate matter being transformed in our system of production. Advances in gene technology have enabled us to begin the engineering and commodificaton of the over 100,000 genes of the human body as well as the genetic makeup of all other living things. With current genetic engineering technology, it is becoming possible to snip, insert, recombine, rearrange, edit, program, and produce genetic material in much the same way as our ancestors were able to separate, collect, utilize, and exploit inanimate materials. Lord Ritchie-Calder, the British science writer, observes, "Just as we have manipulated plastics and metals, we are now manufacturing living materials. . . ."[2]

134

Author Jeremy Rifkin suggests that "Our ultimate goal is to rival the growth curve of the Industrial Age by producing living material at a tempo far exceeding nature's own time frame and then converting that living material into an economic cornucopia."[3]

Without question, genetic engineering represents the ultimate tool in the manipulation of life, the ultimate technology of the human body shop. It extends humanity's reach over the forces of nature, and over the human body, as no other technology in history has. Scientists have become capable of reprogramming the genetic codes of living things to suit our society's social and economic needs. With this newfound ability to manipulate and engineer the genes of living organisms, we assume a new role in the natural scheme. For the first time in history we have the potential for becoming the architects of life itself, the co-directors of evolution. The question of whether we should embark on the long journey in which we become the designers and sellers of "the blueprints" of life is among the most important ever to face humanity.

Though the gene revolution is only a few decades old, its beginnings have already taken on near mythic proportions. The year was 1953, the place was Cambridge, England. There, two young scientists, James Watson and Francis Crick—utilizing X rays, molecular model building, remarkable guesswork, and the accumulated research of many other scientists—discovered the physical makeup of deoxyribonucleic acid (DNA), the fundamental molecule of life. What they found was "a beautiful new structure," a long, twisting double helix located along the chromosomes of a cell. Connecting the strands of this double helix in stepladder fashion are base units. The units, composed of just four different chemical nucleotides—adenine, thymine, guanine, and cytosine—arrange themselves in endless variations of patterns that form the genes. The interaction of genes appears to determine the growth and development of living organisms. Bring a few hundred genes together and you have a simple lifeform, such as bacteria. Combine over 100,000 genes in a certain way and you have a more complex organism, the human body.

On April 25, 1953, Watson and Crick published their historic find in the form of a one-thousand-word article in the British science magazine *Nature*. However humble the presentation, the announcement transformed the field of biology. The British Nobelist Dr. Peter

B. Medawar, whose work had been key in making organ transplants possible, hailed the discovery of DNA as "the greatest achievement of science in the twentieth century."[4] Another observer called it "one of the epic discoveries in the history of scientific thought."[5]

Though stunning, the discovery of the structure of genetic material still left biologists a long way from being able to study and engineer genes effectively, much less initiate a biotechnology revolution in the recombination, redesigning, and production of various genes. In order to splice, delete, and recombine genes into useful products, and then produce them in industrial quantities, two vital steps had to be taken. The first requirement was the ability to chop up DNA into manageable specimens for study, recombination, and cloning. To accomplish this task, some new tool that would act as a "molecular scalpel" needed to be discovered. After years of experimentation researchers, including Drs. Matthew Meselson and Robert Yuan, found in nature the tools they were looking for. These scalpels, an amazing group of enzymes called "restriction enzymes," have the ability to slice DNA as required. The enzymes are naturally produced by bacteria. Nature appears to have designed them to defend the bacteria against invading viruses. The restriction enzymes accomplish their task by cutting up and destroying viruses or bacteria that are attempting to invade their host bacteria.

Once having discovered restriction enzymes and their function, genetic researchers were able to harness the capabilities of these natural scalpels and use them as primary tools in cutting up the immensely long DNA molecules into patterns of smaller pieces. Through the fortunate fact that bacteria produce restriction enzymes, engineers now had the ability to cut DNA into analyzable units of a single gene or a few genes. Had restriction enzymes not occurred naturally, it would have taken decades for researchers to design genetic "scalpels" from scratch.[6]

However, the use of restriction enzymes to isolate genetic material was only the first step toward the "biofacturing" of genes. Once genetic material was cut up and isolated, the next step was to be able to make unlimited copies of the isolated genetic material. In July 1973, twenty years after the discovery of DNA, a research team led by Drs. Stanley Cohen and Annie C. Y. Chang performed an experiment every bit as epochal as Watson's and Crick's discovery of DNA. These researchers inserted genes from *Xenopus laevis*, the South African clawed toad, into a common bacterium called *e. coli*. *E. Coli* uses cell division to reproduce itself and takes only twenty minutes to divide into two identical daughter cells; these in turn divide into four cells, then eight cells, then sixteen cells, and on in a geometric progression, until after a short time millions of copies of the cell are made. Cohen, Chang, and their gene-splicing team were intent on harnessing the multiplicative process of *e. coli* to biofacture identical copies of the toad genes that had been inserted into the bacterium. The researchers found that when the bacterium reproduced itself, it also reproduced the toad genes that had been inserted into it. Through recombining the genes of the toad with the bacterium, the researchers had created a method of biofacturing genes. As the bacterium created millions of copies of itself, it also created exact copies of the genetic material that had been artificially inserted into it. All that remained was for the researcher to harvest the biofactured genes from the bacteria culture medium.[7]

Despite its apparent elegance and ease, the path to biofacturing genes had not been simple. In order for the unique reproduction process, called *cloning*, to work, researchers had to be sure that they inserted genetic material in bacteria in such a way that it multiplied each time the bacteria's own DNA multiplied. After much trial and error, they discovered that a sure and effective way of getting genetic material into the bacteria intact was to use a vector. Researchers attached the desired DNA to vectors like viruses or plasmids (small pieces of DNA carried by certain bacteria) prior to insertion in the bacteria. These vectors carry the genetic material into the bacteria in such a way as to guarantee their replication.[8]

The breakthroughs in discovering restriction enzymes and in the large-scale recombination and cloning of genes made the genetic revolution possible. Now researchers and corporations could isolate valuable genetic material and use cloning and fermentation processes to biofacture these materials at an industrial pace. The engineering techniques that had been used to manufacture products using inanimate material were now being used on the building blocks of life itself.

As the techniques of gene manipulation are refined, the genetic engineers are beginning to create organisms never known before. Microbes, plants, and animals with foreign genes inserted into their permanent genetic code are emerging in increasing numbers from the laboratories of the biotechnology companies. Proponents of this extraordinary new industry promise dazzling benefits. They confidently predict a new biotechnology "green" revolution that will help end world starvation. They prophesy the development of cures for cancer and AIDS. They hail the coming abolition of our most pernicious hereditary diseases.

However, to a generation brought up with naive, techno-booster bromides—"Progress Is Our Middle Name," "Better Living Through Chemistry," "Cheap and Clean Nuclear Power," and even "DDT Is Good for Me"—the promises of utopian biotechnologists ring hollow. Certainly, for a society viewing a new genre of global environmental threats created by industrial pollution, it is now painfully evident that every new technological revolution brings with it both benefits and costs. The more powerful the technology is at expropriating and controlling the forces of nature, the greater the disruption of our society and destruction of the ecosystems that sustain life. Society's experience with both the nuclear and petrochemical revolution bears out this truth.

With genetic engineering we have one of the most powerful and intimate technologies imaginable. And while biotechnology's benefits continue to be heavily advertised, its risks are too little discussed. The full-scale use of biotechnology in agriculture, industry, and in human health and reproduction raises unprecedented ethical, economic, and environmental concerns. For example, the biotechnology industry is preparing to release scores of genetically engineered viruses, bacteria, plants, and animals into the environment. In coming decades hundreds, perhaps thousands, of genetically engineered life-forms may enter the world's ecosystems in massive commercial volume. A central question must be answered prior to any large-scale releases of biotechnology organisms into the environment: What risks do such products pose to human health and the earth's ecology? Because they are alive, genetically engineered products are inherently more unpredictable than chemical products. Genetically engineered products can reproduce, mutate, and migrate.

Once released, it is virtually impossible to recall these living products back to the laboratory. A survey of one hundred top scientists in the United States acknowledged the potential benefits of genetic engineering, but warned that "its imprudent or careless use . . . could lead to irreversible, devastating damage to the ecology."[9]

The use of biotechnology also creates the potential for considerable social and economic dislocation, especially in farming communities. Even a single biotechnological product can have significant adverse impacts. A timely illustration is the recent research and development of bovine growth hormone (BGH). When injected into cattle on a daily basis, this hormone, cloned through genetic engineering, could increase milk production by up to 30 percent per dairy cow. Because of the already flooded milk market, BGH poses a serious threat to dairy farmers. It has been estimated that milk prices may fall 10 percent to 15 percent within the first three years of the introduction of BGH. It has been further estimated that the number of dairy farmers may have to be reduced 25 percent to 30 percent to restore market equilibrium.[10] These economic dislocations, and similar problems resulting from other biotechnology products, will have dramatic social, economic, and cultural effects unless the dissemination of genetically engineered products is strictly controlled both nationally and internationally.

Of greatest concern are the profound and difficult ethical questions raised as genetic engineering vastly expands the human body shop. As transfusion and transplantation technology created the commodification of blood, organs, and other body parts, and as reprotech created the growing market for sperm, eggs, embryos, the born, and the unborn, genetic engineering has begun commodifying the tens of thousands of genes that make up the human genome. We have begun the process of commercially exploiting our common genetic heritage.

A vital element in the expanding genetic marketplace is the ongoing work in mapping, sequencing, and deciphering human genes. At first only a very few human genes were isolated and analyzed for their function in the human body. However, as noted in chapter 9, the United States has now undertaken a $3 billion research program, the Human Genome Project (until recently under the guidance of James Watson), which is an attempt to map and decipher the entire human genome. In 1992, the U.S. government will spend over $160 million on this research project.[11] The Human Genome Project is the largest coordinated effort in biology ever directed to a single goal. But what is the project's goal? Is it knowledge about our genetic makeup, or is it rapid commercial exploitation of genes? Scientist themselves are unsure. The human gene mapping enterprise has been brought to a near standstill due to an unseemly rush of scientist-entrepreneurs who are attempting to be the first to market, patent, and otherwise commercially exploit valuable genes as they are located and mapped. One federal researcher applied for patents on the chemical "tags" of over two thousand brain genes, which he had recently located through his work for the Human Genome Project.[12] Of course, once a valuable gene is located and isolated, it can be produced in huge quantities (through the techniques of splicing and cloning described previously) and sold as therapy for disease, or for any other marketable purpose. Within a few years, large biotechnology corporations could have a monopoly on the ownership and use of the over 100,000 genes that comprise our human genetic makeup.

Genetic engineering also allows scientists to cross species boundaries with ever-increasing ease. This includes the insertion of valuable human genes into animals in order to create better food sources, to fashion more reliable research animals, and even to have genetically engineered animals act as factories to produce valuable human body materials. These techniques go far beyond any traditional methods of breeding animal or plant species. Many feel that cross-species genetic transfers threaten the diversity and genetic integrity of the biotic community. As we will see, prolonged and expanded use of these cross-species engineering feats could mean a significant alteration of the natural world as we now know it.

Moreover, the U.S. Patent and Trademark Office recently granted the first patent on a genetically engineered animal. This regulatory edict, the first commercial patent on any animal, reduced genetically engineered animals, including those containing human genes, to the status of manufactured products. This decision bodes ill for the future. Will succeeding generations of children grow up in a world where the genetic codes of plants, animals, and humans are interchangeable, and living things are patented as engineered products with no greater intrinsic value than mechanical or chemical products?

Biotechnology is not just being used to engineer permanent changes in the genetic code of animals, but also to alter genes in humans. The last few years have seen several experiments involving engineering the genes of human patients. Although these early attempts at human genetic engineering have focused on disease, there are theoretically no limits to the kinds of human genetic traits that could be altered and manipulated. Scientists are currently discussing taking the next step in human gene engineering: altering a person's germline, the genes responsible for passing on hereditary traits. They argue that the time has come for science, through biotechnology, to be a coauthor in the evolution of the human body.

Finally, biotechnology is not only altering life-forms, it is also changing the very nature of reproduction. Scientists are not simply cloning genes, they have discovered ways of cloning mammals— making Xerox copies of rabbits, sheep, and cows. Some now predict that within the next generation we will see the first cloned humans.

———

As policymakers and the society at large struggle with the environmental and economic issues presented by biotechnology, it is becoming clear that the manipulation and marketing of human genes represents the final denouement of the human body shop controversy. As will be seen in the next several chapters, the resolution of the epochal questions raised by the commodification of human genes will determine whether the body is gift or property, whether we see ourselves as inviolable human beings or merchandise.

11

Each new power won by man is a power over men as well.
Each advance leaves him weaker as well as stronger.
C. S. Lewis[1]

A Discriminating Drug

THE IMAGE ON the cover of *Sports Illustrated* was haunting: former all-pro football player Lyle Alzado in close-up. His face was no longer the fierce visage that had sparked fear in so many opposing quarterbacks, but rather a gaunt, emaciated face, with unexpectedly frightened eyes. The impressive head, once perennially helmeted for battle, was now covered with a bandanna to hide the hair loss caused by chemotherapy.[2]

In the magazine's cover story, "I Lied," the former National Football League (NFL) star was going public. After decades of denial, he was now admitting to massive use of steroids and genetically engineered human growth hormone, drugs that he believed caused him to contract inoperable brain cancer.[3]

The story Alzado told to reporter Shelly Smith was both tragic and frightening, a tale of self-destruction endemic to so many American athletes caught in the fiercely competitive world of high school, college, and professional sports. Alzado was a high school all-American football player, but always felt he might not have the "size" to fulfill his dream of playing in the NFL. By 1969, when he was attending Yankton College in South Dakota, Alzado realized, "I wasn't even big enough for a small school like that, so I started taking steroids." His steroid use continued and increased throughout his violent and successful career. Alzado remembers his motivation: "I was so wild about winning. It's all I cared about, winning, winning."[4]

And Alzado was not alone. According to the former player, many of his peers, equally determined to win at all costs, were also using steroids: "No matter what an athlete tells you, I don't care who, don't believe them if they tell you these substances aren't widely used. Ninety percent of the athletes I know are on the stuff. We're not born to be 280 or 300 pounds or jump 30 feet."[5]

Things started to fall apart for Alzado after his retirement in 1985. For one, he kept on taking steroids. "I couldn't stand the thought of being weak." Then, after a few years of "hard to take" inaction, Alzado decided to make a comeback. Comebacks for forty-one-year-olds in the NFL are virtually unheard of. Alzado, though remarkably confident given his long odds, still felt he needed something special to propel his body into shape. He began taking genetically engineered human growth hormone.

Human growth hormone (hGH) is produced by the pituitary gland. It has broad effects on the body. During childhood and adolescence, it stimulates muscle and bone growth and contributes to the development of internal organs and the immune system. In the early 1980s, genetic engineers were able to isolate and clone massive amounts of the hormone using new biotechnology techniques. Human growth hormone was among the earliest human biochemicals mass-marketed by the biotechnology industry. The hormone is believed to have a variety of body-building effects, including its only legal use, increasing height in pituitary dwarfs. Some athletes believe that the hormone helps build up muscles and strength. The ready availability of genetically engineered hGH on the black market has made it the drug of choice for many athletes. And hGH has an additional advantage: Unlike steroids, its presence in the body cannot be detected by drug tests.

Even after Alzado's ill-considered comeback was aborted by a knee injury, he stayed on hGH, though his black-market hGH habit was costing thousands of dollars per month. A short time later, after reporting recurring dizzy spells, double vision, loss of coordination, and slurred speech, Alzado was hospitalized. The doctors could find nothing. Finally, after continued symptoms and a second hospitalization, Alzado was told that he was suffering from a devastating form of brain cancer that is extremely rare, T-cell lymphoma.

Both Alzado and his physician, Dr. Robert Huizinga, an internist practicing in Beverly Hills, California, firmly believed that his cancer was caused by his steroid and hGH use. Huizinga states, "I think there's no question. We know anabolic steroids have cancer forming ability. We know that growth hormones have cancer-growing ability."[6] Huizinga is extremely concerned about the cancer risk for all young athletes on steroids or hGH: "I think we have a real time bomb on our hands."[7] Alzado concluded his story for *Sports Illustrated* with hard-earned advice for all athletes. "If you're on steroids or human growth hormone, stop. I should have." Alzado continued his public education campaign on the dangers posed by hGH and steroids until his death in May 1992.

The hGH time bomb feared by Alzado and Huizinga may be much larger and more explosive than anyone thought. Black market use among male teenagers attempting to "pump up" is soaring. A March 1992 poll found that 5 percent of suburban tenth-grade boys surveyed stated that they used genetically engineered human growth hormone.[8] But muscle-builders and athletes are not the only victims of hGH. Like most Americans, Alzado and his doctor probably didn't know that the genetically engineered hormone they came to fear is being used on thousands of U.S. children every day. These young people are not using this drug at their own choosing, nor are these children ill. They are not athletes fighting for lucrative contracts or body builders intent on championship bodies. The parents of these children are not obtaining the genetically engineered drug on the black market as did Alzado. It is being prescribed by family doctors, even though the use of hGH on many of these children is just as illegal as its use by NFL players or youthful body builders. These children are being subjected to the extraordinary physical and psychological risks of daily genetically engineered hormone injections for only one reason: Their parents feel that they are too short.

The Eye of the Beholder

"Everybody at school calls me shrimp and stuff like that," reports eleven-year-old Marco Oriti. "I feel like a loser. I feel like I'm nothing." Marco is four feet, one inch tall, which is about four inches below the generally accepted average for his age. He weighs forty-nine pounds. He is a serious student and a good soccer player. Despite these achievements Marco is hurt by the taunts of his peers. He dreams of a dramatic turnaround on his tormentors: "One day I want to, like, surprise them. Just come in and be taller than them."[9]

Oriti's parents are doing something about his problem. Marco's mother, Luisa, injects her son with genetically engineered hGH every night but Sunday. Marco has been injected with the drug for six years; and if he continues use for four more years, as scheduled, the cost to the Oritis will be in excess of $150,000.

The Oritis' motives in purchasing the costly drug and in subjecting their son to hGH injections are not dissimilar from the "win at all costs" philosophy that led Alzado down the hGH track. Marco lives in Concord, a city in northern California described as a "high achievement community," a place where "competition begins early." This get-ahead environment provides Luisa and her husband Anthony, a bank vice-president, with the rationale for their use of hGH on Marco: "You want to give your child the edge no matter what," says Luisa. "I think you'd do just about anything."[10] Marco also dreams of getting ahead, including imagining himself playing in the NFL. Whatever his dreams, Marco is unequivocal about the hGH shots: "I hate them."

Of course even without hGH treatment, Marco's expected height leaves many dreams open. As noted by *New York Times* reporter Barry Werth, who first reported Marco's story: "Without hGH, Marco's predicted height was 5 feet 4 inches, about the same as Nobel Prize winning economist Milton Friedman and this year's [1991] Master's golf champion Ian Woosnam, and an inch taller than the basketball guard Mugsy Bogues of the Charlotte Hornets."[11]

Human growth hormone has not been approved for use in children like Marco any more than it has for athletes like Alzado. As noted, the drug's only approved use is for the treatment of a rare congenital disease, pituitary dwarfism, which results when the body is unable to manufacture its own growth hormone. About 2,500 Americans suffer from this disorder. But through a combination of willing doctors and a full court press by Genentech—the chief U.S. manufacturer of hGH and a company known for its aggressive marketing—hGH is being injected into thousands of kids just like Marco. These children have no disease or disability; they are simply below average height.

What makes the use of hGH by families like the Oritis all the more disturbing is that there is no proof that the drug actually increases ultimate height for children like Marco who do not have hormone deficiency. In fact, many pediatricians are openly pessimistic about the drug's efficacy in improving height for short children. As UCLA endocrinologist Dr. Douglas Fraser states, "I'm less convinced as time goes by that there's any long-term benefit here." Dr. Selna Kaplan, a consultant for Genentech on hGH, has even speculated that hGH treatment may speed up the process of puberty, thereby causing children to stop growing sooner than they would normally.[12] The National Institutes of Health (NIH) is also ambivalent:

> The long-term effect on adult height of growth hormone in children who have some growth hormone of their own has not yet been established. It is possible that it will have no effect, since growth hormone may cause earlier closure of bone growth plate, so that the same adult height would be reached sooner.[13]

Moreover, once you start injections, you're hooked. You must continue shots, as with Marco, for years, lest you actually retard a child's growth. As stated by the NIH,

> Stopping growth hormone treatment in children who are not growth hormone deficient before they have reached their adult height may cause them to grow more slowly than they did before treatment. This may occur because taking the extra growth hormone causes the body to temporarily stop making its own growth hormone.[14]

There is also no evidence that use of the hormone will allevi-ate feelings of inferiority or depression relating to height. In fact, many experts believe hGH treatments may do more psychological harm than good. As reported in the *Journal of the American Medi-cal Association:*

> There is currently no evidence that hGH treatment alleviates psy-chosocial morbidity in non-hGH deficient short children. . . . The use of hGH for an otherwise normal child might itself result in psychosocial morbidity, in the form of stigmatization, reinforcing self-perception as abnormal, and subsequent loss of self-esteem. If hGH fails to increase height, both the child and parents may suffer from feelings of failure.[15]

The Leukemia Link

As the Alzado experience indicates, there are serious potential health risks to the children being given hGH. By the late 1980s, scientific studies were suggesting a possible causal rela-tionship between long-term hGH use and leukemia. In February 1988 Japanese researchers reported that their research "strongly suggests an association between the development of acute leukemia and long-term hGH administration in hypopituitary patients."[16] This report was shortly followed by a similar finding by European researchers. These researchers, led by Professor J. C. Job and several colleagues from the hGH working group of the European Society for Pediatric Endocrinol-ogy, surveyed hGH treatment programs in seventeen countries in Europe and discovered six hGH-treated patients who had developed leukemia during or after treatment.[17]

These troubling findings prompted the meeting of the Inter-national Workshop on Growth Hormone and Leukemia to examine the relationship between hGH and leukemia. This workshop con-cluded that in certain circumstances the apparent rate of leukemia in hGH patients increased by a factor of three over those who did not receive treatment.[18] This report, while inconclusive, did note that

"There is a possible increase of leukemia in GH-deficient patients treated with hGH."[19]

The U.S. Food and Drug Administration (FDA) is responsible for approving drugs like hGH and regulating their use. This is not a comforting thought for those hoping for stricter regulation of hGH. Throughout the last several years, the FDA has been rocked by scandals and exposés on its inefficiency in regulating various drugs. Recent problems for the agency have surfaced in several areas, from generic drugs to off-label uses of a wide variety of pharmaceuticals. Former FDA chief Frank Young, forced out of office by the mounting controversy over his agency's actions, was an unapologetic booster of deregulation and a tireless promoter of biotechnology.

However, even the FDA has been carefully looking at the hGH leukemia connection. This is due in part to the fact that the agency's own reporting on hGH indicates that the hormone may be causing a variety of health problems for hundreds, even thousands of children. The FDA maintains a readout system on approved drugs that provides preliminary information on potential side effects associated with or related to treatment with a certain drug. The FDA's 1991 report on hGH drugs documents hundreds of cases of side effects that may be related to hGH use. These reactions included several cases of leukemia as well as other serious health concerns, including hyperthyroid and respiratory problems. All in all, the FDA has reported that twenty-seven cases of leukemia have occurred in those being treated with hGH, though the agency claims that many of the cases may be coincidental to hGH use.[20]

The manufacturers of hGH have noted the growing leukemia concern connected with their product. In 1989, both Genentech and Eli Lilly changed their labeling on hGH to include information that leukemia had been reported in a "small number" of children who have been treated with hGH. Additionally, in response to legal pressure, the FDA has announced that it is considering changing its labeling requirement for hGH so that it "more fully describe[s] the available information on leukemia associated with growth hormone administration. . . ."[21]

As information on hGH's health effects adds to its other drawbacks, the picture for children like Marco becomes profoundly troubling:

- Such children are incapable of informed consent.

- They are involved in a therapy that means hundreds of injections each year.

- It will cost more than $150,000 over a ten-year period.

- Their treatment is not for any disease, but rather is based on the fact that they are perceived as short.

- The treatment has not even proven effective to increase height in these children.

- They are hooked on the drug for years, since stopping treatment may actually retard their growth.

- Experts believe that far from helping, the treatment itself could be psychologically harmful.

- The children could be under an increased risk of contracting leukemia and suffering other serious side effects.

Despite this dismal list of risks and disadvantages, the use of hGH by parents for short children appears to have been a gold mine for Genentech and Eli Lilly (the other major U.S. producer of hGH). The U.S. market for the drug is now well over $200 million. In 1991, Genentech sold over $185 million worth of Protropin (the brand name for their hGH product), an increase in sales of over 60 percent in three years. By the end of 1991, hGH had become Genentech's leading product. Few seriously believe that this volume of sales, and significant increase in use, is attributable solely to hGH's approved use on dwarfs. Genentech attributes the climb in hGH sales to increased use of the hormone among young people and an increase in the diagnosis of growth hormone deficiency in young people.[22]

How is this possible? With all its extraordinary and potentially tragic drawbacks, how could Genentech and Lilly be selling hundreds of millions of dollars worth of their hormone? How could parents and physicians be seduced into treating their children with this hormone? The sad truth is that, as a result of societal prejudice, the carelessness of physicians, and the unscrupulous behavior of drug companies, these children have become victims of a cure searching for a disease. They are targets of biotech and pharmaceutical companies seeking profits for their human body commodity.

A Cure in Search of a Disease

Prior to the advent of genetic engineering, human growth hormone was extracted from human pituitary glands taken from cadavers. Coroners from around the country would remove the small pituitary glands from the dead and ship them to laboratories, where technicians would remove the growth hormone from the glands. This relatively primitive method of production just barely kept supply of the hormone up with demand. However, since demand was limited, there was no urgency in obtaining new sources of growth hormone.

As the early biotech companies began developing genetically engineered human growth hormone, the general view was that it would not be a major profit-making product due to the fact that it could only be marketed to the relatively small number of patients suffering from dwarfism. This view underscored the ability of Genentech's product Protropin to qualify as one of the first so-called "orphan drugs." Orphan drug status is intended to reward companies that develop drugs to treat uncommon, relatively unprofitable health problems. The status allows the company a guaranteed monopoly on the drug for seven years. Eli Lilly has also received this status for its version of hGH.

In the early 1980s, even as Genentech and Lilly pleaded orphan status for their allegedly unprofitable drug, there were rumblings that the companies had bigger plans for hGH. Some industry analysts were predicting a $100 million a year market in the United States for hGH.[23] This prediction, made three years before hGH approval, confused many biotech analysts. They cast about for answers as to how the market for this little-needed pharmaceutical would reach these heights. In 1983, author Edward Yoxen proved prophetic when he wrote,

> It has been suggested . . . that cheapening growth hormone might lead to its abuse by people wanting to be tall but who are not pathologically short. Knowing that height is a source of anxiety to many people, particularly adolescents, it seems possible that unscrupulous suppliers might seek to market height augmenting drugs.[24]

A 1984 Hastings Center Report noted that a "promotional campaign" for the drug had already begun prior to its availability. The writers warned that, "Since growth hormone is likely to be profitable, it will certainly be aggressively marketed."[25]

In 1985, producers of genetically engineered hGH got their chance. That summer doctors discovered that four people who had received natural hGH had died of Creutzfeldt-Jacob disease, a rare dementia cause by slow-acting viruses that infect the brain. The FDA, alarmed that there appeared to be a dangerous viral contaminant in natural human growth hormone, promptly removed all natural hGH from the market. Genentech was nearing completion of its investigative trials of Protropin at the time of the natural hormone recall, and by the end of the year had received FDA approval to market its genetically engineered hGH. While in prior years there had been just barely enough hGH to go around, genetic engineering allowed for hGH to be produced in massive quantities. As noted with concern by Dr. James M. Tanner of the Institute of Child Health at the University of London, "We are now moving from an era in which there were too many patients chasing too little growth hormone to an era in which there will be too much growth hormone chasing too few patients. It is really Brave New World."[26] Genetically engineered growth hormone was an expensive cure searching for patients.

The two companies manufacturing hGH responded to the marketing challenge. Through relentless promotion of hGH, they essentially created a new disease for which they had the cure: being short could now be seen as a disease. Many advocated the idea that a child who was in the lower 3 percent of height for his or her age group was no longer considered well. The short child needed treatment. Creating this new group of drug consumers showed economic acumen. Of the 3 million children born each year in the United States, 90,000 will by definition be in the bottom 3 percent in height. Experts calculate that this could be an $8 billion to $10 billion market. But that's only half of it. The use of the lowest three percentiles in height as a definition of disease need never stop. If hGH actually does work, and makes shorter children taller, the growth curve will still always have a bottom 3 percent. As noted by reporter Werth, "Someone will always be shortest." There will be an endlessly self-perpetuating market for hGH.[27]

The aggressive marketing of hGH has not received rave notices from ethicists. "Until growth hormone came along, no one called normal shortness a disease," says Dr. John D. Lantos of the Center for Clinical Medical Ethics at the University of Chicago. "It's become a disease only because a manipulation [hGH] has become available and because doctors and insurance companies, in order to rationalize their actions, have had to perceive it as one. What we're seeing is two things—the commodization of drugs that are well-being enhancers and the creeping definition of what it means to be healthy."[28]

Others, including the federal government, were more supportive of this new "creeping" redefinition of shortness as a disease. In 1988, the National Institutes of Health (NIH) initiated a twelve-year study of the effects of hGH on otherwise normal "short" children. The research project, under the supervision of its principal investigator, Dr. Gordon B. Cutler, is using dozens of children to test the effectiveness of Lilly's version of hGH, Humatrope.[29] In the past Lilly has given significant funding to NIH for testing of Humatrope. This support has included payment for patient travel expenses.[30] The company also has freed NIH from any liability should the experiments cause harm to children receiving the hormone.[31] Lilly is hoping to use the experiments—research in part paid for by the federal government—as a basis for getting permission from the FDA to legally expand the commercial marketing of hGH to healthy short children like Marco.

The children in Cutler's project, girls and boys starting generally at age nine or ten, are subject to hGH injections three times a week for up to ten years.[32] Regular checkups include a variety of extensive physicals, precheckup ingestion of hormone pills or an injection of testosterone, X rays, blood tests, various psychological tests, nude photographs against a height grid (the NIH notes that "the child's eyes will be covered in the picture if they wish"), and other similarly intrusive procedures. One estimate is that the growth experiments are costing the U.S. taxpayer approximately $200,000 per child. The children also are exposed to the many side effects of hGH use, including the possible link to leukemia.

The NIH-Lilly hGH program appears to violate not only accepted norms on experimenting on children but also the agency's own regulations which limit research on children to cases where there is serious disease. However, the NIH is unapologetic about its research program. NIH spokesperson Micheala Richardson states, "These kids are not normal. They are short in a society that looks at that unfavorably."[33]

Public interest and medical groups have, over the years, petitioned the NIH to halt the hGH program. In their pleas to the NIH, concerned groups have cited the health risks to the children involved in the research and also pointed out that the NIH is treating a non-disease shortness as if it were a disease. Until recently, the NIH ignored the compelling arguments for halting its hGH program. However, in August 1992, responding to yet another legal action taken by two public interest groups and increased media attention on its controversial program, the agency agreed to convene a "monitoring body" to assess the ethics and legality of its program. The NIH also committed itself to stopping enrollment in the program until its study had been completed.[34]

"Too Short"

The invention of a new "disease" is in itself ethically disturbing. However, there is an even deeper and more alarming aspect to the promotion and use of hGH—a concern that goes beyond the usual endemic greed and exploitation of the medical marketplace. Human growth hormone treatment is not, after all, just another cosmetic product or procedure sold to the youthful American health consumer, such as ear piercing or acne treatment. Unlike these other practices, hGH use exacerbates a very real prejudice in our culture, a prejudice some call heightism.

A growing amount of evidence indicates that height may affect key aspects of becoming successful in our society. Job prospects, salary level, even chances of winning political office may be affected by one's height. Employment studies generally show that taller job candidates do better than their shorter counterparts. One recent survey has shown that a surprising 72 percent of recruiters admit to preferring taller candidates over shorter competitors. Another survey compared the salary differences between male library science graduates who ranged between six feet, one inch and six feet, three inches, and those under six feet, with the salary differences between those who were in the top half of their class academically and those in the lower half. The average difference in starting salary between the taller and shorter graduates was more than three times greater than the difference between the more and less academically qualified.[35]

Once an individual has been hired, height may also affect promotion and raises. Using a sample of over five thousand men who had passed an Air Force cadet qualifying exam, a study revealed that after twenty-five years those who were five feet, six inches to five feet, seven inches tall were earning significantly less than those who were six feet or over. It has also been well reported that height plays an important role in political races. Short candidates have been known to stand on risers behind podiums in order to make their TV appearance taller. But heightism goes back a lot further than TV. Since 1904, the taller candidate for president of the United States has been the victor in 80 percent of the elections, whereas Republicans and Democrats can only claim about a 50 percent rate. In fact, of all U.S. presidents, only two, James Madison and Benjamin Harrison, have been shorter than average for an American male at the time they were elected.[36]

Human growth hormone treatment of "normal" short children like Marco reifies and exploits heightism. It terms as disease what is simply a product of prejudice. "Short stature is, to some extent, a natural variation, and the associated psychosocial morbidity results from cultural prejudice," note authors of a recent article in the *Journal of the American Medical Association*. "We do not usually call prejudice induced conditions, which confer cultural disadvantages but have no intrinsic negative health effects, disease."[37] Some are more direct. "It's warped" says Diane Keaton, of the National Association of Short Adults—which is described as a "semi-serious lobby group whose motto is 'Down in Front'"—"We think shorter is better." She notes that short has distinct ecological advantages: "We use less food and fiber. We take up less space."[38]

Behind Genentech's marketing strategy, and the NIH experimentation on those that our society "looks at unfavorably," is an insidious and potentially profound reversal of social burden. The implicit suggestion behind the marketing of hGH, and its use on short children, is that the way to handle prejudice is to physically alter its victims. We will have truly entered a brave new world if we continue creating designer drugs like hGH, or other genetically engineered therapies, intended to remove physical characteristics or traits that are the target of prejudice. Should we view being African-American as a disease to be treated with genetically engineered pigmentation therapy to alter skin color? Should we see being female as a disease to be treated by embryonic sex-change surgery? It is worrisome enough that we are screening fetuses for genetic predispositions to various nondisease traits such as gender and aborting on that basis, as was described earlier. We cannot add to those eugenic practices the culturally enforced medicalization of victims of prejudice. The solution to prejudice is education of those who are its purveyors, not the genetic engineering of their victims.

Flouting the Law

It came as a relief to many hGH critics when, in 1990, Congress passed a law making it a felony to distribute hGH for unapproved uses. The amendment to the Food, Drug, and Cosmetic Act seems clear: "Whoever knowingly distributes, or possesses with intent to distribute, human growth hormone for any use in humans other than disease or other recognized medical condition . . . is guilty of an offense punishable by not more than 5 years in prison. . . ."[39]

In practice, however, a combination of ambiguous science and regulatory boondoggles make it extremely unlikely that this provision will be adequately enforced in cases like Marco's. First of all, determining who is growth-hormone deficient, and therefore legally able to receive the drug, is far from easy. Most short children are not deficient in growth hormone. Their height is usually attributable to genetic factors or constitutional delay. Often the two factors come together in the same child, a child like Marco, and this may lead to a marked shortness of stature without any hormonal deficiency. As has been noted by many pediatricians, finding out which of these short children may actually be hormone deficient is at best an "unreliable" business. The difficulty in diagnosing hormone deficiency has been more of a boon than a bother to Genentech. Because of this difficulty in diagnosis, the company can blur the line between the majority of short children who are not hGH deficient, and should not receive the drug, and the small minority who are. "I can take you in to one endocrinologist who'll say your kid is hormone deficient and another who'll say he isn't," claims Genentech's chief medical officer, Dr. Barry Sherman. "It's a total crapshoot."[40] Genentech has effectively spread this pro-hGH analysis to pediatricians across the country, often through lavish hGH symposiums. Endocrinologist Douglas Fraser states the case bluntly, "Genentech co-opted the field."[41]

Even in clear cases of inappropriate use of hGH, Genentech is unlikely to find itself criminalized. Increasingly, companies are taking advantage of lax FDA enforcement and making huge profits on nonapproved uses of new drugs. In fact, it is well known that drug companies often play a kind of regulatory shell game, finding the narrowest and easiest uses of a drug for FDA approval and then reaping massive dividends on off-label uses.

Despite the growing controversy over their drug, the manufacturers of hGH are pushing ahead for official approval of hGH use for short-stature children. They also have come up with a novel disease for their cure: aging. Natural secretion of hGH slows and even stops in most men after the age of fifty. Many would like to call this natural reduction in hGH secretion a disease. In the last two years, the manufacturers of hGH have sponsored numerous trials of the drug on men over sixty-one years of age to see if it helped reverse natural aging processes like muscle shrinkage and accumulation of fat. In 1990, as the first published study seemed to show that the hormone did have some effect on reducing aging, "gee whiz" stories appeared in the nation's dailies on the new genetically engineered "fountain of youth." Business analysts speculated that millions of Americans might be added to the hGH market, as it was not only marketed to the nation's children but also to its aged. The early euphoria was somewhat dampened by subsequent analysis tending to show that the desired effect of the drug reversed itself when the drug was no longer being taken. Additionally, side effects include swollen joints, enlarged breasts in men, and carpal tunnel syndrome. Researchers also worry that the hormone's ability to spur cell growth may promote cancers, especially breast cancer in women.[42]

The marketing of hGH is a cautionary tale about the mass-marketing of human substances produced by genetic engineering. If the current indiscriminate sale of hGH is an accurate omen, we can expect increasing exploitation by the biotech industry of social prejudice, targeting the victims of such prejudice for expensive pharmaceutical or genetic makeovers. And unfortunately, we can expect increased suffering for those who, like Lyle Alzado and Marco Oriti, are caught up in a society that is too often propelled by competition and profit rather than compassion and tolerance.

———

The current marketing and use of human growth hormone demonstrates the hazards that accompany the expansion of the human body shop to include genetically engineered human hormones and other biochemicals. But genetically engineered drugs and pharmaceuticals have only a limited ability to alter human physiology and psychology. From the beginning, the genetic engineers have had a more ambitious vision—the use of biotechnology to alter the genetic makeup of human beings, and perhaps to become co-architects of human evolution itself.

12

Modern genetics is on the verge of some truly fantastic ways of "improving" the human race . . . but in what direction?
James B. Nagle[1]

Engineering Ourselves

IN AN AGE of protests, this was the first of its kind. It was early March 1977, and hundreds of demonstrators had flocked to the futuristic, domed auditorium of the National Academy of Sciences (NAS). The protesters chanted slogans such as "We will not be cloned," and they carried signs bearing warnings, including "Don't Tread on My Genes."

The object of the protest was a three-day symposium being held under the auspices of the NAS. The forum was intended to bring together scientists, government officials, and business leaders to discuss the future prospects of genetically altering life-forms, including humans. The chairman of the meeting, Dr. David Hamburg, president of the NAS Institute for Medicine, undoubtedly had anticipated that this would be the usual scientific conference, a collegial discussion of current scientific and legislative issues that had been cropping up as a result of advances in genetic manipulation. It was not to be.

The demonstrators, led by activist Jeremy Rifkin, crowded the auditorium with their signs and dominated the session with their chants and shouted questions to the symposium's panels. They relentlessly prodded the scientists and bureaucrats, urging them to confront the moral and ethical implications of engineering the genetic code of life. They also repeatedly demanded that speakers disclose who was financing their research. (The forum was supported in part by funds from a variety of drug manufacturers.) Finally, under a barrage of questions about the eugenic and discriminatory potential of biotechnology, the chairman had no choice but to offer the podium to Rifkin and others to air their concerns.

Speaking up with the protesters were many prominent scientists. At a press conference prior to the demonstration, Nobel Prize winner George Wald called the use of genetic engineering "the biggest break in nature that has occurred in human history." Renowned biochemist Dr. Erwin Chargoff warned against the use of genetic research to attempt to control the evolution of humans and other life-forms.[2]

The activists and scientists who voiced their concerns that day were part of a growing chorus of those who feared the engineering of life. As early as 1967, Marshall Nirenberg, the Nobelist who first described the "language" of the genetic code, had delivered a stern lecture about engineering human beings, along with a remarkably prescient prophecy:[3]

> My guess is that cells will be programmed with synthetic messages within 25 years. . . . The point that deserves special emphasis is that man may be able to program his own cells long before he will be able to assess adequately the long-term consequences of such alterations, long before he will be able to formulate goals, and long before he can resolve the ethical and moral problems which will be raised.[3]

The fears of the early gene engineering critics focused on proposals to engineer the human germline—to permanently alter the genetic makeup of an individual that is passed on to succeeding generations. Many scientists were predicting that, by manipulating the genes in sperm, eggs, or embryos, future physicians would be able to excise "bad" genes from the human gene pool. Critics envisioned a future human body shop industry in eliminating the genes responsible for sickle-cell anemia or cystic fibrosis by mass engineering of these "problem" genes from the sex cells (the sperm and ova) of individuals. Future genetic engineers could also add foreign genes to a patient's genome, genes from other humans or even different species. These genes might protect an individual from various diseases, or confer desired qualities like better looks or brains. Ultimately, they believed that as scientists learned more about the relationship of genes to disease and other human traits, there would be an inevitable push to treat life-forms as so many machines whose working parts, genes, could be engineered or replaced if they were "defective."

Moreover, it was clear that if the genetic engineering of human beings should come, and most believed it would, there would be a quantum leap in both negative and positive eugenics. No longer would it be necessary to attempt to carefully control generations of breeding to create "good" characteristics, or to resort to sterilization, abortion, or genocide in order to remove abnormal or undesirable traits. Individuals could be altered through genetic surgery that would repair or replace bad genes and add good ones. Nobel Prize winner Jean Rostand's early visions of the eugenic potential of gene engineering went even further: "It would be no more than a game for the 'man farming biologist' to change the subject's sex, the colour of his eyes, the general proportions of body and limbs and perhaps the facial features."[4] Many agreed with scientists such as Wald and Chargoff that the genetic alteration of people could eventually change the course of evolution. In 1972, ethicist Dr. Leon Kass wrote, "The new technologies for human engineering may well be 'the transition to a wholly new path of evolution.' They may therefore mark the end of *human* life as we and all other humans know it."[5]

For over two decades, scientists, activists, ethicists, and the media have engaged in the debate over the medical and moral questions surrounding the germline genetic engineering of human beings. Editorials have appeared with headlines questioning "Whether to Make Perfect Humans"[6] and how to arrive at "The Rules for Reshaping Life."[7] Many critics have continued to argue against the entire enterprise of "the remaking of man." They question the wisdom of having scientists decide which part of the human genome should be eliminated and which enhanced. And if not scientists, who, they ask, will determine which human genes are bad and which good? They warn that even supposedly "bad" genes may bring extraordinary benefits to humanity. Recently, it was discovered that cystic fibrosis genes appear to provide individuals with protection from melanoma, an increasingly common form of skin cancer. Research conducted in the 1980s determined that sickle-cell anemia genes appear to help provide individuals with immunity to malaria. Excising such genes from the human gene pool in the effort to eliminate human disease could backfire with potentially catastrophic results.

There is also the question of how and when society will ensure that the powerful technology of germline gene engineering will be limited to the treatment of serious human diseases. As described in prior chapters, genetic screening of embryos is already being used for eugenic purposes, including sex selection; and genetically engineered drugs are being used for cosmetic purposes in a way that helps foster certain forms of discrimination. Who will ensure that germline therapy is not abused in the same discriminatory and eugenic way? Will those with under normal height or I.Q. become key targets of the future entrepreneurs of germline therapy? Other novel legal questions arise from the prospect of germline therapy, issues similar to those being asked in reference to advances in prenatal genetic screening. Do children have the right to an unmanipulated germline? Or, conversely, do they have a right to the best germline that genetic surgery can offer and money can buy?

As the debate around germline gene therapy continues, another form of human genetic engineering has already begun. This form of genetic manipulation does not involve sex cells, but rather those cells that do not partake in reproduction. These cells are called *somatic cells*. Engineering these cells is both easier and far less controversial than attempting to manipulate germ cells. Altering somatic cells triggers far less concern about eugenics, in that the cells being repaired or added affect only the single individual being engineered. They do not affect the inheritance of genetic traits. Early uses of somatic cell engineering include providing individuals with healthy or repaired genes that might replace those that are faulty and causing disease.

Though somatic cell gene therapy does not affect the genetic inheritance of future generations, there are still fears. Will individuals with "poor" genetic readouts—those predisposed to a variety of disorders or abnormal traits—be under pressure by parents, education providers, insurance companies, and employers to undergo gene therapy to remove their "bad" genes? Will the therapy be used "cosmetically" to add or eliminate nondisease traits, such as growth, skin color, or intelligence? Will victims of discrimination be pressured by societal prejudice to alter in themselves those traits society views as negative?

The early concerns about germline and somatic cell genetic engineering relied primarily on future projections of the potential abuse of the technology. However, two early cases involving misuse of gene therapy contributed significantly to the controversy that marked the early years of experimentation on the genetic manipulation of humans. The first scandal involving the nascent technology happened over two decades ago.

False Start

Between 1970 and 1973, German physician G. H. Terhaggen, with the assistance of an American scientist, Dr. Stanfield Rogers, injected Shope papilloma virus (SPV) into two sisters (eighteen and five years of age) and one other infant girl. All three patients were suffering from an extremely rare disease known as hyperargininemia. The disease is caused by a genetic defect in the enzyme whose job it is to break down the amino acid arginine. Without the enzyme to break it down, arginine builds up in the bloodstream, resulting in epileptic seizures, spastic paraplegia, and severe mental retardation. The two sisters were hopelessly ill at the time of the injections. The experiment offered them no hope of a cure.

Rogers was familiar with SPV and had observed that researchers working with the virus had unusually low levels of arginine in their blood. From this Rogers surmised that SPV might break down arginine in the blood system and be useful in treating SPV. Rogers defended injecting the virus into children, describing SPV as a "passenger" virus that does no harm to humans. His description was not wholly accurate. Earlier experiments had shown that cancer could be triggered in rabbits when they were given high doses of SPV.

News reports about the Terhaggen-Rogers procedures caused a storm of concern among the public, scientists, and Congress. Rogers was accused by some of unethically using the children, reducing them to the status of research animals. Ethicist Paul Ramsey wrote, "[The experiments] are a case of genetic *experimentation* in which men have chosen retardates and taken from them something which they cannot give, namely free consent to prolong their dying for the sake of scrutinizing it scientifically."[8]

The Terhaggen-Rogers experiments were unsuccessful, and no other SPV experiments for treating hyperargininemia have been reported since. The Rogers case did have several lasting impacts. The scandal surrounding the case was a key incentive for the first proposed legislation on genetic engineering. Senator Walter Mondale, after holding hearings on the issue, introduced a bill for a "National Commission on Health, Science and Society" to investigate and study the legal, social, and ethical implications of genetic research. Though it received some support, the bill was never voted on.[9]

In 1976, in the wake of continuing concern evoked by the Rogers case and continuing inaction by Congress, the National Institutes of Health (NIH) took the initiative in attempting to regulate human gene engineering. The NIH, usurping the role that Congress had eschewed, formed the Recombinant DNA Advisory Committee (RAC) to promulgate guidelines for federally funded DNA research, including any use of human gene engineering. The guidelines, which were first published in 1976, required that any experiment involving human gene therapy be approved by an applicant's university or hospital institutional review board prior to seeking final approval of the RAC itself. The RAC guidelines were regularly amended over the next several years and remain the central regulatory authority in the genetic manipulation of human beings. It was little noted at the time that the NIH undertook the ambitious task of regulating gene research despite the fact that the agency was and still is the main actor and funder of DNA research. Over the last several years, this conflict of interest has led inevitably to a number of "fox-guarding-the-henhouse" scenarios, as RAC scientists discuss the propriety and scientific merit of gene experiments or procedures in which they may have a significant financial or personal interest.

Seven years after Rogers completed his controversial gene transfers, another scandal hit the gene technology world. Between May and July 1980, two young women—one in Italy, the other in Israel—unwittingly became pioneer patients in the world's first sophisticated experiments involving the genetic engineering of humans. Dr. Martin Cline, a scientist at the University of California, was a leading researcher in the genetic engineering of animals. In April 1980, Cline successfully engineered a foreign gene into a number of mice. The gene endowed the mice with resistance to a certain drug. Only a few weeks later, emboldened by his success, Cline was unable to resist making an ill-considered leap from mice to people as he experimented on the two women.

Cline's two patients suffered from thalassemia, an inherited genetic defect in the hemoglobin gene that impairs normal production of red blood cells. Mild cases result in anemia; more severe cases cause death in early childhood. Cline had submitted a proposal to his local review board for injecting genetically engineered cells into patients to treat certain blood disorders, including thalassemia. However, in July 1980 the local review board disapproved the protocol, agreeing with outside consultants that the procedure had not been proven effective or safe. Cline made the decision to go abroad to do the experiments, regardless of NIH procedures or approvals. Unfortunately, he neglected to tell his patients or host countries about the true nature of his human genetic engineering experiments.

The procedure was not a medical success, and it was an ethical disaster. Cline violated the NIH guidelines and numerous professional norms. He deceived his two patients, and he evaded scientific review of his experiment. Cline was subsequently reprimanded and punished by the NIH.

As with the Rogers case before it, the Cline experiment contributed to a surge of legislative and regulatory action on human gene engineering. This included a 1980 "blue ribbon" White House Commission, which had the unwieldy name "The President's Commission for the Study of Ethical Problems in Medicine and Biomedical and Behavioral Research." Its executive director was ethicist Alexander Morgan Capron. In November 1982, the commission issued its report on the ethical implications of human gene technology. The report, called *Splicing Life*, was generally enthusiastic about technology and science, calling current advances a "celebration of human creativity."[10] The report did underscore many concerns about gene engineering, and even alluded to germline therapy as "eugenics."[11] However, it failed to call for restrictions or significant limitations on germline human therapy.

The report did not receive a uniformly favorable response. Criticism of the report was especially acerbic in regard to its recommendation of a ban on human-animal hybrids (ridiculed by some as a ban on "mermaids"), research that few were seriously proposing, while maintaining a cautionary wait-and-see attitude toward germline engineering of human traits, which many were advocating. As stated in a *New York Times* editorial:

> The commission has boldly vetoed animal-human hybrids, which are on no one's agenda, but has tiptoed around more concrete issues. Let mermaids be free of bans, at least until someone has serious plans of creating them. The more tangible problem is whether to permit heritable changes to the human gene set.[12]

Despite its ambiguous tone and uneven reasoning, the report was historic. It was the first issued by a government body on the subject of the genetic manipulation of humans.

Concern over the Cline cases and recommendations contained in the White House Report had other results. In 1982, the U.S. Congress held the first comprehensive hearings on human gene therapy. By 1984, the congressional Office of Technology Assessment (OTA) prepared and issued a report on the subject. Additionally, the NIH acted to shore up its regulation of human gene work by establishing a Working Group on Human Gene Therapy. The Working Group, later to become the Human Gene Therapy Subcommittee, was to play an important role in reviewing early human gene therapy research projects and procedures. Finally, in 1985, Congress authorized the establishment of a Biomedical Ethics Review Board, whose mandate included the examination of the ethical implications of the genetic engineering of humans, and the proposing of legislative limits on the uses of gene technology.

Throughout the 1980s, the criticisms of gene therapy continued. In 1983, Jeremy Rifkin organized a religious and scientific coalition against the use of genetic engineering on humans. The coalition and its signed statement opposing germline engineering were front-page stories around the United States. Unlike Capron's commission, the coalition's resolution on germline therapy was unambiguous: "Resolved, that efforts to engineer specific genetic traits into the germline of the human species should not be attempted." Its logic on prohibiting heritable gene alterations was also straightforward: "No individual, group of individuals, or institutions can legitimately claim the right or authority to make such decision on behalf of the rest of the species alive today or for future generations."[13] The resolution, which was presented to Congress, was signed by a remarkable variety of religious leaders, including mainstream Jewish, Catholic, and Protestant religious organizations, as well as by many prominent scientists.

Six years later, an important and detailed religious statement on biotechnology was issued by the World Council of Churches (WCC). It contained a strong policy statement calling on all churches to support a "ban on experiments involving the genetic engineering of the human germline."[14] The WCC was also deeply concerned about somatic cell gene experiments. The report called upon member churches to urge "strict control on experiments involving genetically engineered somatic cells, drawing attention to the potential misuse of . . . [this technique] against those held to be 'defective.'"[15] The timing of the WCC statement could not have been more pertinent, for 1989 was to be the year that the age of human genetic engineering officially began.

Playing God?

On January 30, 1989, almost twelve years after the first demonstration on human genetic engineering, another such protest took place. The protesters came to a meeting of the National Institutes of Health Recombinant DNA Advisory Committee (RAC). Since publishing its guidelines in 1976, RAC had met dozens of times to discuss and approve experiments in genetic engineering. The advisory committee, composed mainly of scientists, held meetings that were usually staid affairs replete with lengthy discussion of arcane data and procedures.

This RAC meeting was like no other. There, demanding to be heard by the NIH scientists and genetic engineers, were fifteen of the nation's most prominent leaders in disability rights, many themselves suffering from disabilities. Additionally, several biotechnology activists were present to demand accountability of the scientists on the RAC. Many of the scientists appeared visibly uncomfortable at the prospect of discussing human gene engineering with people concerned about a new age of eugenics—and all under the unaccustomed glare of TV cameras. Those present knew that they were at a historic moment in the genetic engineering revolution, for this RAC meeting had as an agenda item discussion of approval for the world's first legally sanctioned genetic engineering experiment on humans.

The experiment involved genetic engineering but was not intended to be a cure. Researchers wished to insert novel genetic "markers" into certain immune cells taken from the bodies of terminally ill cancer patients, and then transfuse those cells back into the patients. With the help of the markers, they hoped to track which cells were working effectively and which were not. The procedure was to be carried out by the NIH's prime genetic engineering team of Drs. French Anderson, Steven A. Rosenberg, and Michael Blaese.

Minutes after RAC chairman Dr. Gerard J. McGarrity called the meeting to order, critics began to express deep concern that the NIH had begun the historic process of approving human gene engineering protocols while still doing nothing to put in place a review process on the ethical and legal implications of human genetic alteration. Jeremy Rifkin announced that his Foundation on Economic Trends had filed suit that morning, calling on a federal court to halt the experiment until the NIH committed itself to allowing the public a greater voice in decision on gene therapy. Rifkin also noted that the lawsuit was based on the fact that the historic experiment was approved by a secret mail ballot, the first in RAC's history. He repeated the concerns he and other demonstrators had expressed over a decade before: "Genetic engineering raises unparalleled ethical and social questions for the human race. They cannot be ignored by the NIH. If we are not careful we will find ourselves in a world where the disabled, minorities, and workers will be genetically engineered."[16] Another protesting voice at the meeting was Evan Kemp, then commissioner of the Equal Employment Opportunity Commission (EEOC), and himself disabled:

> The terror and risk that genetic engineering holds for those of us with disabilities are well grounded in recent events. . . . Our society seems to have an aversion to those who are physically and mentally different. Genetic engineering could lead to the elimination of the rich diversity in our peoples. It is a real and frightening threat.[17]

Those present asked the RAC to set up an outside review board for human genetic engineering experiments that would include experts in the rights of minorities, workers, and the disabled. They insisted that the RAC scientists, though astute on advances in genetics, were no experts in the public policy implications of their work. "This group cannot play God when it comes to deciding what genes should be engineered in and out of individual patients," Rifkin said during heated arguments with members of the committee. "What will be the criteria for good or bad genes? Who will decide what genes, and which people, will be engineered?" he continued. "The people in this room are just not qualified to raise these monumental social issues. You're just not going to be able to maintain that control of power within a small group. We need to broaden this group."[18] A few members of the RAC board became belligerent, denying, sometimes angrily, the suggestion that they lacked the expertise to oversee the larger social and political implications of their work. Others simply ignored the proposal. When the vote came, the RAC board unanimously (twenty in favor, three abstentions) turned down the proposal to set up a public policy review committee.

The RAC critics lost the NIH vote, but they won the battle in court. On May 6, the NIH settled the law case filed against the NIH, agreeing to immediately make changes in the RAC guidelines that would forbid mail or secret ballots and would also provide more review for gene therapy experiments.[19] The legal settlement cleared the way for the first legally sanctioned gene engineering experiment on humans. The gene "marker" experiment took place a few days later, on May 22, 1989.

Claiming Immunity

The second gene experiment on humans was performed just over a year after the first. It was the first official attempt to use somatic cell human gene engineering as a therapy for disease. On September 14, 1990, a four-year-old girl from Cleveland with the immune disorder popularly known as the "bubble boy syndrome" was injected with a billion cells into which a new gene had been inserted. The girl was born without the gene that controls successful functioning of certain immune cells called T lymphocytes. The rare condition (it affects only about twenty children worldwide), known as adenosine deaminase (ADA) deficiency, leaves victims helpless in the face of disease and infection. Many children suffering from ADA deficiency have been kept alive by isolating them in a germ-free capsule, as was "David," the famous "Boy in the Bubble" at Baylor College of Medicine in Houston, Texas.

Dr. French Anderson and a team at NIH intravenously infused the child with blood cells containing the missing ADA gene in hope that it would help her recover normal functioning of her immune system. On the surface the medical procedure looked little different from a normal blood transfusion. The procedure, which took place in the Pediatric Intensive Care Unit of the Clinical Center of NIH, in Bethesda, Maryland, lasted twenty-eight minutes. One hour later the young patient was wandering around the hospital playroom, eating M&Ms.[20]

The young girl who had become the first human gene therapy patient to be legally engineered with human genes became something of a celebrity, as did Dr. Anderson. The media reported the historic occasion in glowing terms. Soon reporters were writing about "Dr. Anderson's Gene Machine."[21] After some initial reports of success, it was not uncommon to hear that genetic engineering had cured the "bubble boy syndrome." A second patient began gene treatment in January 1991. It was hard to imagine a more altruistic beginning for a technological development that so many had feared as the beginning of a new eugenic movement.

The experiment had its dark side, however, including some unfortunate parallels with Rogers's scandalous experiments on children in the early 1970s. A careful examination revealed that Anderson's procedure may have been more hype than cure. The "bubble boy syndrome" cases were now a misnomer: None of the handful of existent cases required the bubble to protect the immunologically impaired children from disease. Since the mid-1980s, these children were being adequately treated with a new drug therapy. Anderson, however, had started his research into ADA before the drug therapy was available. Many felt that he continued on with his protocol more out of stubbornness and ambition than medical necessity. Months before the experiment took place, members of the Human Genome Subcommittee had openly questioned Anderson on the rationale for subjecting children to the risks of gene therapy when they were already being treated successfully. So concerned were the RAC members about the effectiveness of Anderson's therapy that they restricted Anderson and his team to working only with patients who were already receiving the drug therapy. This in turn led to the question of how Anderson could accurately assess the results of his experiment. One scientist noted that it would be a little like attempting to assess the results of aspirin on a patient who was being treated with antibiotics.

Whether or not Anderson is using his patients as gene therapy guinea pigs, his experiments appear to violate the general bioethical rule that the expected benefits to an individual from an experimental therapy should equal or exceed the potential harm. The experiment's protocol was clear. The procedure did not offer children suffering from the genetic disorder a cure, but merely a supplemental therapy. The beneficial results of the experiment are at best marginal. A cure awaits improvements in bone marrow transplantation.

By contrast, the dangers to children from Anderson's experiment could be quite real. Anderson and others involved in inserting genes into patients use animal retroviruses to carry those genes. The retrovirus used in all early gene therapy experiments, including the ADA experiment, is one called murine leukemia virus (MuLV). It is a retrovirus obtained from mice. Anderson engineers the ADA gene into the retrovirus and then injects the gene package into a patient. Once inside the patient, the retrovirus invades cells and drops off the genes. Genetic engineers like Anderson attempt to render these carrier retroviruses harmless, but there are still concerns that these viruses could cause cancer or other serious disease in patients. Except in the case of Anderson's ADA experiments, MuLV had only been approved for use in terminally ill patients in whom the retrovirus could do little additional harm. Yet Anderson used this suspect retrovirus on children who were living relatively normal lives with potentially long life spans ahead of them.

In December 1991, less than a year after Anderson began genetically engineering his second patient, an unsettling report was made public. A researcher, Arthur Nienhaus, described his discovery that the MuLV virus had caused cancer in primates. The researcher suspects that the cancer may have been caused by a contaminant that leaked into the virus during production.[22] Anderson and others were quick to note that they used a different system to produce their MuLV, one less prone to contamination. However, the discovery bolsters the view that much more needs to be learned before MuLV is widely used as a gene therapy tool.

In a rare demonstration of scientific breaking of ranks, several fellow genetic engineers openly expressed their displeasure with the Anderson experiment. One gene therapy expert called the Anderson procedures "absolutely crazy." Dr. Arthur Bank, professor of medicine and human genetics at Columbia University, charged that gene therapy researchers at NIH were driven by ambition and not by good science. "The main impetus [for the ADA experiment] is the need for French Anderson to be the first to do gene therapy in man. . . . This may turn out to be bad news for all of us," Bank told a genetics conference within a week after the experiment had started.[23] Dr. Stuart Orkin, professor of pediatric medicine at Harvard Medical School, noted, "A large number of scientists believe the experiment is not well founded scientifically. . . . I'm quite surprised that there hasn't been more of an outcry against the experiment by scientists who are completely objective."[24] Dr. Richard Mulligan, a pioneer in gene therapy work and a member of the RAC board—the only one who voted against the experiment—was more direct. "If I had a daughter, no way I'd let her get near these guys if she had that defect."[25]

Anderson has more than his experiments to defend. Critics of the approvals of the first gene therapy experiments also point out that over a five-year period, Anderson has almost singlehandedly pioneered delivering federally funded human gene engineering research to a private company with which he is a collaborator. In 1987, Anderson did what many viewed as "scientifically unthinkable" when he joined forces with venture capitalist Wallace Steinberg to help build a human gene engineering company, Genetics Therapy, Inc. (GTI), a company one observer has called the "ultimate body shop."[26]

Steinberg had long headed the venture capital arm of Johnson & Johnson and was looking for a new market challenge in what promised to be the cutting-edge industry of the future—human genetic engineering. Traditionally, government scientists have regarded joining forces with private investors as unseemly if not unethical. Anderson's relationship to human gene engineering entrepreneurs has cast a shadow over both the science and the procedures that led to the approval of the first of several human gene therapy experiments. Concerns about conflict of interest were heightened in late 1990 when GTI hired former NIH/RAC chairman Gerard McGarrity. McGarrity had been a leading supporter of GTI's and Anderson's gene therapy experiments, and as chairman of RAC had helped shepherd the therapy proposals through the NIH approval process. In 1991, GTI's numerous maneuvers paid off: Sandoz Pharma, Ltd., one of the world's major multinational companies, bought $10 million of GTI stock and agreed to provide $13.5 million over the subsequent three years in project funding. GTI ended 1991 with cash and marketable securities of $20.8 million.[27]

Human gene engineering is progressing quickly. Currently, over a dozen somatic cell gene engineering experiments are ongoing on three continents.[28] Numerous other gene engineering protocols are being developed for approval in the near future. Large-scale use of gene engineering to cure disease or cosmetically change individuals is still several years away; nevertheless, the scandal, ambition, and moral blindness that have characterized the early history of human genetic engineering set a profoundly disturbing precedent for the future.

Moreover, many of the protections against abuses in the use of gene technology put in place in the 1980s are fast disappearing. The Congressional Biomedical Ethics Board, established in 1985, was disbanded in 1990. Additionally, in 1991 Dr. Anderson and others successfully urged the disbanding of the RAC Human Gene Therapy Subcommittee. Finally, in the face of a massive influx of profit-seeking and potential conflicts of interest, the viability of RAC as a responsible regulatory agency of human gene engineering is in considerable doubt.

In the future we will be genetically engineering ourselves in numerous ways—applications of biotechnology with which our society is ill prepared to deal. As researchers successfully locate genes responsible for height, weight, and I.Q., there are still no restrictions that would prevent an industry from altering these traits through somatic gene therapy. Further, researchers are now more determined than ever to begin the first germline gene engineering experiments on humans. There is general consensus that such research will become a reality over the next decade. We have no national or international mechanisms that will prevent germline engineering from permanently altering our human genome, no restrictions on the unlimited genetic alteration of sperm and eggs, or the engineering of embryos. Despite continuing controversy, publicity, and massive public funding of gene technology research, the questions demonstrators shouted at scientists over fifteen years ago have still not been answered.

———

As the genetic engineers edge closer to the ultimate human body shop activity—the germline engineering of humans—they are already engaged in wholesale engineering of the germlines of the rest of the biotic community, including other mammals. While we still hesitate on the brink of permanently altering our own genetic traits, we are already engineering the genetic makeup of the rest of the living kingdom. We are even engineering human genetic material into animals. Perhaps in seeing what genetic technology is doing to the rest of the animal kingdom, we will get a glimpse of what the future holds for the genetic engineering of ourselves.

13

*What people lose sight of is that genetic engineering is
about engineering—taking the principles of technology and
applying them directly to the genetic code of microbes,
plants, animals and humans. . . . It involves the application
of criteria of utility and efficiency to all these life forms.
The potential benefits are marvelous—but in the end the
philosophical assault on our concept of life is at least as
impressive.*
       Jeremy Rifkin[1]

The Beast Machines

THE HUMAN IMAGINATION has never been limited
by species boundaries. The Greeks imagined a mythological creature,
the Chimera, a hideous, fire-breathing she-monster with the body of
a goat, the head of a lion, and the tail of a dragon. The Chimera,
sacred beast to the goddess Artemis, represented darkness, drought,
and the underworld, and its name has become the generic term for
all such mixed-species creatures. Other ancient chimeras include the
Griffin, Greek symbol of enlightenment and superior mind, with its
body and legs of a lion and beak and wings of an eagle; the Hindu
Garuda, the half-man, half-bird destroyer of evil, and means of transportation for the great god Vishnu; and perhaps the most famous
chimera, the Egyptian Sphinx, with its human head representing
wisdom and its lion body symbolizing strength.

The chimeras of old glorified the "spirits" believed to inhabit
the living world. The shapes and characteristics of animal bodies
were seen as embodiments of spiritual and emotional qualities. For
the ancients, the creation of chimeras was a mixing of these qualities

to create meaningful and lasting symbols of eternal essences—wisdom, strength, evil, stubbornness, obedience, or beauty. Chimeras were the marriage of diverse animal bodies and souls to represent aspects of the "world-soul."

In recent years biotechnologists have resurrected the practice of creating chimeras that transcend species boundaries. But this time the chimeras are not solely figments of human imagination. They are real-life *transgenic* animals—animals engineered to contain the genetic traits of humans and other species. Unlike the mythological creatures of past ages, these chimeras are not icons with religious or sacred meaning; rather, they represent attempts by genetic engineers to create more efficient and profitable animals for the food and medical marketplaces. Researchers are engineering human genes, and those of other species, into livestock and poultry in order to create "super" animals for slaughter and consumption. Human genes are also being inserted into research animals to make them more valuable research tools in the laboratory. And some animals are even being genetically engineered to function as biological factories for the production of valuable human body materials, including insulin, hemoglobin, and blood-clotting agents. Together, these modern-day chimeras represent the newest and among the most bizarre manifestations of the human body shop.

Five-ton Cows and Twelve-foot Pigs

Pig No. 6707 was meant to be "super"—super fast-growing, super big, super meat quality. He was supposed to be a technological breakthrough in animal husbandry—among the first of a series of high-tech animals that would revolutionize agriculture. Researcher Dr. Vernon Pursel and his colleagues at the United States Department of Agriculture (USDA) research center in Beltsville, Maryland, had created this pig to be like no other—and, to a certain extent, they had succeeded. No. 6707 was unique, both in his physiology and in the very core of each and every cell. For this pig was born with a human gene engineered into his permanent genetic makeup.

Using microinjection techniques, Pursel and his team had placed a human gene governing growth into the pig while it was still an embryo. The idea was to have the human growth gene become part of the pig's genetic code and thus create a pig that would grow larger and faster than any before. Injecting human genes into the fertilized eggs of swine, and then replanting the engineered embryos into surrogate mother pigs for gestation, is far from an easy or efficient process. Piercing embryonic cell walls with a needle that injects the human growth gene destroys a high percentage of the embryos being manipulated. In fact, in several years of experiments, scientists injected more than eight thousand embryos with genetically engineered growth genes to produce just forty-three transgenic animals.[2] But No. 6707 beat the odds, becoming among the first pigs in history to produce human growth hormone from his pituitary gland.

With his transgenic creation, Pursel had hoped to mimic the heralded results achieved by fellow genetic engineer Dr. Ralph Brinster. In 1982, Brinster, working out of the School of Veterinary Medicine at the University of Pennsylvania, electrified the scientific community with his successful creation of a "super" mouse, a chimera mouse engineered to contain the human growth gene as part of its permanent germline. Science journals featured the startling picture of two female mouse siblings side by side. The one without the human gene weighed twenty-eight grams; the mouse with the gene dwarfed its sibling and weighed fifty-nine grams, over twice as much.[3]

Researchers assumed that what worked for mice would work for livestock. They dreamed of future farms with pigs and other livestock growing many times larger than current animals. However, Pursel's transgenic pig did not turn into a super pig. The human genetic material they had injected into the animal altered the pig's metabolism in an unpredictable and unfortunate way. The mixture of genes had proved too complex to effectively control, and No. 6707 had turned into a tragicomic creation. Excessively hairy, lethargic, riddled with arthritis, apparently impotent, and slightly cross-eyed, the pig could hardly stand up. The USDA tried to rationalize its experiments by noting that although the transgenic pigs were not bigger, their large muscle mass would make their meat "leaner."

Despite whatever spin Pursel and others tried to put on their transgenic creations, many viewed No. 6707 as the wretched product of a science without ethics.

Ever since the prospect of genetic engineering was thrust on the world in 1973, with Stanley Cohen's and Herbert Boyer's breakthrough experiment in inserting foreign genes into a living organism, scientists have been alarmed about the potential powers that biotechnology placed in the hands of engineers. The ability to alter the very blueprint of life seemed more like science fiction than science fact. For many in the scientific and political community, mixing and matching the genetic makeup of the entire biotic community to create "made-to-order" animals and plants was part daydream and part nightmare.

It took only a few years for pioneering researchers, such as Brinster and Pursel, to turn these techno-dreams into engineered reality. Now, each year, researchers around the world are conducting tens of thousands of experiments in order to create transgenic animals. In 1990 alone, researchers in Great Britain inserted foreign genes into 11,399 pig embryos, creating sixty-seven transgenic animals. Bioengineer Dr. John Clark of the Institute of Animal Physiology and Genetic Research at Edinburgh, Scotland, admitted that many of his pigs, like their American counterparts, continued to suffer stress and joint problems. But the researcher vowed to continue his work in the name of "growth and feed efficiency."[4] Elsewhere, British researchers have also genetically engineered 4,500 sheep embryos, resulting in thirty-four transgenic animals.[5] Researchers in Sydney, Australia, inserted the insect-killing gene from a tobacco plant into the sweat glands of sheep in the hope that when fly larvae burrow into the sheep's skin they would be killed by the enzyme produced by the engineered genetic material.[6] Scientists in West Vancouver, Canada, have experimented with boosting salmon size with chicken and cattle growth hormones.[7]

As the creation of transgenic animals increases internationally, the United States remains the undisputed leader in the genetic engineering of animals. Over the last decade, American government and private researchers have spent billions of dollars, much of it provided by taxpayers, in the creation of a variety of transgenic animals. Dr. Pursel and dozens of other scientists have inserted a wide range of genes, including over two dozen different human genes, into animals and plants. Rodents and livestock containing human genes have become commonplace at several U.S. corporate, university, and government laboratories. Carp, catfish, and trout have also been engineered with a number of genes from humans, cattle, and rats to increase their growth and increase reproduction.[8] Plants have received "antifreeze" genes garnered from flounder to help them grow at lower temperatures. Following highly publicized English experiments, researchers at the University of California at Davis used cell-fusion techniques to create "geeps," astonishing goat-sheep combinations with the faces and horns of goats and the bodies of sheep.

Some transgenic creations seem more the stuff of late-night comedy shows than serious science. Researchers from the University of California at San Diego took the fluorescent genes from fireflies and engineered them into tobacco plants to create plants that "glow in the dark." They succeeded in creating tobacco plants that light up twenty-four hours a day. Tongue-in-cheek speculation on the rationale for this experiment included the idea that it was conducted in the interest of helping inveterate smokers find their packs of cigarettes even in the dark.[9]

Despite the initial repulsion and even humor accompanying the sight of many transgenic animals and plants, the new technology represents an epic step in the history of engineering life and the commercialization of human biological material. The engineering of animals with genetic material from humans and other species, all in the name of productivity, utility, and profit, is an astonishing advance of the ideology of efficiency into the very heart of living matter. In prior centuries, our control over the reproduction of plants and animals was limited. People could only guess at the laws of heredity and could breed plants and animals only very slowly, over generations, to attain certain characteristics considered useful in agriculture and animal husbandry. Sophisticated techniques for cattle breeding were not developed in England until the middle of the nineteenth century. Gregor Mendel's theories on genetics were only resurrected at the turn of the twentieth century. Even with the growth in the science of genetics, until quite recently the human ability to intervene in creating animal characteristics was still constrained by the limits of species interbreeding, the internal controls of natural reproduction insuring that the genetic integrity of creatures would withstand human intervention.

Now, with the development of genetic engineering, the species boundary is broken. Scientists using gene insertion techniques—microinjection, cell fusion, electroporation, retroviral transformation—are increasingly able to transfer genes across species boundaries. There are still some limits. Up to this point, transgenic manipulation has involved the transfer of only a single gene, such as that triggering growth, between species. However, most believe it is only a matter of a few years before scientists will be able to engineer multigene traits into plants and animals. Recently, the congressional Office of Technology Assessment (OTA) predicted that within the next decade we could witness the engineering of complex genetic traits, including those involving human behavior, into other species.[10]

When contemplating the controlling power of cross-species genetic engineering, many scientists display the "gee whiz" demeanor that has become an endemic part of modern technological development. Researcher J. Mintz, who successfully transplanted rabbit growth genes into mice, predicts that in the future we will see "five ton cows, and pigs twelve feet long and five feet tall."[11] Others have predicted the creation of a wide variety of transgenic creatures—from monster chickens, to oysters that will survive in polluted waters, to wolves that avoid sheep. "Right now, we don't know what the limits are," said Michael Phillips of the OTA. "All the traditional rules we thought about the . . . animal kingdom . . . are thrown out the window."[12] "Are we tinkering with nature? There's no question," exclaims Iowa State University genetics researcher Walter Fehr. "Man has been doing that since we've been civilized. But he's never had this power before."[13]

Other genetic engineers are more cautious about prospects for the near future. USDA researcher Pursel finds a mechanistic metaphor to describe his view of where the research currently is: "We're at the Wright Brothers stage compared to the 747. We're going to crash and burn for a number of years and not get very far off the ground for a while."[14] Of course, it is animals like Pursel's Pig No. 6707 that pay the price for the researcher's "crash and burn" persistence.

From the very first, some people have seen the historic issues created by cross-species genetic engineering. Then-Senator Albert Gore, Jr. noted that "People understand at a gut level that there is something wondrous, and perhaps perilous about a technology that changes the blueprint of life and will force us to make choices that are likely to be more profound than anything we, as a society, have ever faced."[15] Critics have raised fundamental ethical and moral concerns about whether we have the right to use genetic engineering to attempt to redirect the evolutionary process—especially its use in crossing species boundaries. "We are bringing a completely human centered utilitarian attitude toward life," states Dr. Michael Fox, a veterinarian and spokesperson for The Humane Society of the United States. "All of earth's living things will simply become items to exploit."[16] Many commentators have also noted the large number of questions that the technology raises. Is it ethical to alter the genetic codes of animals? Do animals have a right to their own genetic integrity? Is there a limit to the number and type of human genes that should be engineered into other animals, plants, or insects? Who will set appropriate limits to the technology—the U.S. Congress, the president, world leaders, the United Nations, scientists, theologians? "The issues range from ethics within universities, to the environment, to eugenics, to definitions of nature, to religious thought, to what it is to be human," states Dorothy Nelkin, a professor at Cornell University. "Other disputes over technology have been much simpler. . . ."[17]

"Super" AIDS?

Researchers are not altering the germline of animals merely to create cheaper, more abundant, and leaner meat. They are also genetically engineering animals for laboratory use. Engineers have attempted to create valuable research animals by taking human cancer genes as well as genes from other human diseases and disorders and genetically engineering them into the permanent genetic code of animals. These new chimeras often contain lethal human disease genes in every one of their cells. Biotechnologists hope that these animals will be superior research models on which to test important new drugs and therapies. At first glance the creation of chimeras for research appears both altruistic and harmless, but nothing could be further from the truth. The creation of transgenic animals containing human disease genes poses unique human health and environmental hazards, as well as a potential for needless cruelty to animals. A remarkable and timely example is the "AIDS mouse."

From the beginning of the AIDS epidemic, scientists have been searching for the "perfect" research animals for use in their urgent work on the deadly virus. Use of chimpanzees was both controversial and expensive, and though certain primates can be infected by the AIDS virus, it does not make them ill. Normal mice were not good research "tools" because they were not susceptible to infection with AIDS, and therefore did not provide an acceptable human model. Genetic engineer Dr. Malcolm Martin set out to change all that. He began the ambitious attempt to create a transgenic mouse containing the human AIDS virus in its genetic code. At first Martin and his research team at the National Institutes of Health (NIH) in Bethesda, Maryland, transplanted only a piece of the AIDS genetic code into mice. After this initial success, Martin asked the next logical question: "Why not try to duplicate this, by having a copy of the whole genetic code [of HIV] in every single cell of the mouse?"[18]

In the fall of 1987, Martin took the plunge and transplanted the entire AIDS virus genome into lab mice. His team first harvested fertilized eggs (ova) from a female mouse. Then, in a microinjection procedure similar to that used by Pursel and others, Martin's team used a microscopically fine needle to inject the AIDS genome into the nuclei of the fertilized mouse eggs. The AIDS-infected embryos were then put into surrogate-mother mice to gestate. About 10 percent of the offspring carried the AIDS virus as part of their permanent genetic makeup. They carried the virus in every cell of their bodies. Martin's experiments represented the first time that the complete genetic code of an organism causing a lethal disease in humans had been introduced into another animal species.[19]

The first generation of AIDS mice appeared normal. The infected mice did not appear to suffer from AIDS-like symptoms. When they matured the AIDS mice were mated with different strains of mice to see if the AIDS genes would be passed to the next generation. After several tries, the researchers announced their success. Some offspring from one of these matches carried the AIDS virus genome. Unlike their parents, however, this second generation of AIDS mice displayed symptoms associated with human AIDS, such as pneumonia and certain skin conditions.[20] A valuable research model of AIDS seemed to have been created.

The creation of the AIDS mice was hailed as a breakthrough by the media and by many in the scientific community. However, some researchers, even certain scientists at NIH, were disturbed and concerned. There were no regulations to assure the safety of the AIDS mice experiments, and the research proposal for the experiments had not been reviewed or considered by the NIH's Recombinant DNA Advisory Committee (RAC). The central concern of many was that by creating a nonhuman repository for a human disease as volatile and fatal as AIDS, Dr. Martin had created a potential nightmare of extraordinary proportions. If the genetically engineered animals escaped and mated with other mice, there could be an uncontrollable risk of spreading AIDS. Dr. Liebe F. Cavalieri, a scientist at Sloan Kettering Cancer Institute in New York and an expert on biotechnology, said that the experiment represented a serious concern both for himself and others in the scientific community.[21] Sheldon Krimsky, a former RAC member, was prophetic in his view that, "When you begin taking a virus and putting it in a different species, you can't guarantee that the methods of spread and infection will always be the same as they have been."[22] Jeremy Rifkin and the Foundation on Economic Trends immediately filed suit against the NIH, demanding that the agency prepare an environmental and human health assessment on the AIDS mice experiments. Rifkin also expressed grave concerns about extending the host range of HIV into other species.[23]

In the early days of the controversy over the AIDS mice, Martin and other scientists struck back at critics of the experiments. Dr. Philip Leder of the Harvard Medical School, a champion of the creation and commercialization of transgenic mice, termed Rifkin a "cynical" obstructionist and dismissed concerns about the experiment.[24] Dr. Ann Kiessling, also of the Harvard Medical School, rejected the idea that the mice might spread the virus: "I think the chance of the mice actually expressing the whole virus is practically zero."[25] Martin himself dismissed fears about his mice causing the spread of AIDS as "farfetched" and a "*National Enquirer* scenario."[26] Martin noted that his experiments were being conducted under stringent biological containment. He assured the public nothing would go wrong.[27]

A year after the experiment began, disaster struck. According to NIH sources, on Saturday, December 3, 1988, electric power providing air to the AIDS mice was apparently inadvertently shut off by a repairman. When researchers came in early Sunday morning, they found the laboratory without power and only three of the 130 mice alive. Martin was called early Sunday about the accident, while he was still in bed. Commenting later, he called the accident "a pretty major setback for my program."[28] He did not explain how this kind of accident could occur with an experiment that was purportedly among the most carefully monitored in NIH history.

An even greater problem lay ahead for Martin and other researchers involved in creating transgenic mice as human disease models. In February 1990, a paper was published in the respected journal *Science*. The paper reported on experiments conducted by a distinguished team of scientists, including Dr. Robert Gallo, the co-discoverer of the AIDS virus. The research involved investigating the suitability of transgenic mice like Martin's for AIDS research. The findings were extremely disturbing. The Gallo team reported that the use of transgenic mice for AIDS research could lead to HIV interacting and combining with the mice's own "natural" viruses. The study described how this interaction of the AIDS virus with mouse viruses could actually produce new and potentially more dangerous forms of HIV—a "super" AIDS. The paper indicated that these "super" AIDS variants might infect a wider range of cells and might be spread by novel routes.[29] According to commentator Jean Marx, one of these novel routes might be airborne transmission of the new "super" AIDS virus.[30]

Given its findings, the Gallo team had little choice but to recommend renewed caution in creating transgenic mouse models containing the genetic code of AIDS. But safety was not the only concern underscored by the Gallo study. Clearly, if the AIDS virus changes its properties by interacting with native mouse viruses, scientific results obtained from the mouse are not reliable. Results might be at best irrelevant and at worst misleading. According to the study, "Our observations should prompt a critical evaluation of the experimental data obtained from some animal models for HIV-1 and of their suitability for use in the study of new antiviral therapeutic . . . approaches."[31] After several years of federally funded research, Martin's work now seemed fatally flawed. The AIDS mice were not only a new and unique potential danger in spreading AIDS, they also were capable of creating dangerous variants of the AIDS virus. Moreover, they were poor and potentially misleading research models for use in understanding the deadly disease.

The findings of the Gallo team mirrored and vindicated many of the scientific and safety concerns of those who had criticized Martin's experiments as premature and potentially dangerous. The idea that inserting the AIDS virus into the permanent genetic code of mice might enhance the virus's virulence was not a farfetched, tabloid scenario, but rather a scientific probability. The view that the animals could be made into perfect or even useful research "tools" for HIV had been seriously undermined. As with Pursel's experiments, Martin's highly publicized march into the transgenic future had become a disorderly retreat, leaving animal suffering, media hype, false hopes, and large taxpayer expenditures in its wake.

Animals as Biofactories

We have seen how animals have been engineered with human genes to make them more valuable food or research commodities. As scientists continue to engineer animals for these purposes, recent reports from the frontiers of transgenic animal research describe yet another group of chimeras. These transgenic animals may represent the ultimate in the use of animals as part of the human body shop. Researchers are now engineering human genes into animals that force the animals to produce medically useful and valuable human biological materials. They are creating animals that are pharmaceutical factories of human body shop material.

Genetically transforming animals into manufacturing mechanisms for human biochemicals started in the early 1980s, when scientists engineered sheep that created human growth hormone in their blood, and pigs that created human insulin in a like manner. After the human genes, now part of the animal's permanent genetic code, created sufficient amounts of the desired human product, the animal was slaughtered and the valuable human hormone or other biochemical was harvested. This slaughter technique had the obvious disadvantage of requiring the destruction of the animal, the manufacturing plant, prior to retrieving the product. The goose was being killed in order to retrieve the golden egg. So researchers began to search for ways to engineer animals to produce valuable human materials on a sustained basis, and harvest those chemicals, without having to destroy their new animal "factories."

Researchers now think they have found a solution. The new approach involves genetically engineering new breeds of sheep, goats, and cows so that they secrete valuable human pharmaceuticals in their milk. There was early skepticism about this approach, despite the fact that in 1987 researchers successfully engineered a human gene into mice that made the mice produce a human protein in their milk. In September 1991, three independent research teams jolted the biotechnology world by announcing major advances in getting animals to secrete valuable proteins in their milk. "Two years ago, people were doubtful of this technology," noted animal biotechnologist Robert Bremel of the University of Wisconsin. "But now the work shows that the mammary gland can be used as an impressive bioreactor."[32]

What excites Bremel is the prospect of unlimited production of previously scarce and therefore valuable human proteins by the "mammalian bioreactors" in larger milk-producing animals. Among the human proteins being manufactured by the three research teams are blood-clotting factors needed to treat hemophilia patients and heart attack victims, and a protein (AAT) that is being investigated in the treatment of emphysema. Researchers from Pharmaceutical Proteins, Ltd., in Edinburgh, Scotland, and the Agriculture and Food Research Council's Institute of Animal Physiology and Genetics, also in Edinburgh, have genetically engineered sheep that yield milk containing up to thirty-five grams per liter of human AAT. According to the research team leader, Alan Colman, it's "a good start." By late 1991, a team of researchers at Tufts University, along with a team at Genzyme Corporation of Cambridge, Massachusetts, also reported progress. They had engineered goats to produce tissue plasminogen activator (t-PA), a protein used to dissolve blood clots in cardiac patients. Their best goat produces the "clotbuster" at the rate of about three grams per liter of milk.[33]

The recently announced successes do not mean that animal-produced human drugs will be on the market in the near future. Significant technical problems must be solved before animals can be used as bioreactors. Yields of biochemicals from transgenic animals are still not high enough to allow for the use of the animals as profitable commercial production "factories." Moreover, manufacturers will have to show that the new animal-produced human proteins are biologically equivalent to the natural ones and work in the same fashion. Perhaps most important, there are safety concerns. Human proteins made in animals could contain animal viruses that would be harmful to human patients. Sheep and goats are susceptible to scrapie, a degenerative brain disease, and cows are susceptible to the bovine equivalent, bovine spongiform encephalopathy (BSE), known as "mad cow disease."[34] Both diseases might infect humans. Additionally, a significant percentage of cows in the United States have bovine leukemia virus, and about 10 percent of cows suffer from bovine immunodeficiency virus (BIV), a close relative of the AIDS virus.[35]

It is too soon to know whether researchers can, as they advertise they will, overcome these problems and reap major payoffs on transgenic animals as biofactories. Regardless of their ultimate success, however, they already have one distinction. They have, without doubt, produced the most unusual and ethically problematic production system for human body shop materials.

The creation of transgenic animals and their use in the human body shop is profoundly disturbing. The alteration, through genetic engineering, of the permanent genetic code of animals represents a unique and unprecedented assault on their dignity and biological integrity. While the abuse of animals in our factory farms and in our laboratories is widespread and well documented, the threat of transgenic research to the animal kingdom is even more compelling and long-range. If we continue over the next decades to mix and match the genes of the animal kingdom to suit our desires and to gain commercial profits, it could bring about the end of nature as we have known it.

Over the last five years, researchers and corporations creating transgenic animals have gone one step further, adding insult to their injury of the genetic integrity of the animal world. They are now attempting to gain commercial patents on transgenic animals, just as if they were mechanical or chemical inventions. The patenting of life-forms includes not only transgenic animals, but also human genes, cells, and embryos. The patenting of life has become a central issue in the debate over the commercialization of humankind.

14

To justify patenting living organisms, those who seek such patents must argue that life has no "vital" or sacred property. . . . But once this is accomplished, all living material will be reduced to an arrangement of chemicals, or "mere compositions of matter."

> Ted Howard, amicus brief before the U.S. Supreme Court in the Chakrabarty case, 1979[1]

We've known for years that we are a bag of chemicals, mind you a very special bag of chemicals.

> Jeffrey Powell, Associate Professor of Biology, Yale University, at the 16th International Congress of Genetics, 1988[2]

The Patenting of Life

OVER TWO HUNDRED years ago, Thomas Jefferson introduced America's first Patent Act. Jefferson, an amateur scientist himself, was determined to ensure that "ingenuity should receive a liberal encouragement." Jefferson's law was passed in 1793. It provided for inventors to patent "any new and useful art, machine, manufacture or composition of matter, or any new useful improvement [thereof]."[3] As established by this law, a patent is a grant issued by the U.S. government giving the patent holder the sole right to make, use, or sell an invention within the United States during the term of the patent—generally seventeen years. Whatever the "machine, manufacture, or composition of matter" invented—airplane, computer, pesticide, or toaster—the patent provides a government-sanctioned monopoly to the inventor of the product. It is a financial reward for the inventor's ingenuity and is designed to help the inventor recoup the time and expense consumed in creating the "new and useful" invention.[4]

For better or worse, the U.S. economy has been built on innovation rewarded by the patent system. Over 5 million patents have been issued since 1793.[5] Patents provided the profit trigger for the machine age in America. It's an American truism that behind each great discovery lies a patent. Examples are numerous, and notables include the following:

1840: Samuel Morse receives a patent for the telegraph.

1876: Alexander Graham Bell patents the telephone.

1898: Rudolf Diesel patents the internal combustion engine.

1906: Wilbur and Orville Wright receive a patent for their "flying machine."

1938: Vladimir Kosma Zworykin receives a patent for the cathode ray, a forerunner to today's TV picture tubes.

1956: Carl Djerassi patents the birth control pill.[6]

On June 20, 1991, National Institutes of Health (NIH) researcher Craig Venter, prompted by the legal advice of NIH technology transfer attorney Reid Adler, filed a four-hundred-page patent application of a type never imagined by Jefferson, and more startling than any of the millions that preceded it—a patent request that represented a bold expansion of the human body shop and a major development in the industrialization of life. Venter's application requested patents that would result in patent protection and ownership for 337 genes found in the human brain. This controversial application is seen by many in the scientific community as the first step in the patenting of all of the 100,000 or more human genes.[7] A few months after the first application, Venter applied for patents involving two thousand more brain genes. If granted, these patents would give Venter and the NIH ownership of roughly 5 percent of all human genes.[8] In July 1992, Venter announced that he was leaving NIH to start his own gene-mapping institute with the help of $70 million from a venture capital fund.[9] Clearly, locating and patenting genes could become big business.

Venter is one of dozens of scientists around the world involved in gene sequencing and mapping. Like many others taking part in the various human genome research projects, Venter's work is somewhat mechanical. His job is to locate the chemical sequences that make up genes. He and the other researchers have no idea what function the genes they are "tagging" actually perform in the human body—it may take years before their functions in human biology or psychology are even partially understood. However, no matter how deep their current ignorance about the function of the brain genes they have tagged, Venter, Adler, and others at the NIH want to patent them all. Any one of the genes could turn out to be extremely valuable, perhaps a key gene for brain cancer research or future therapies to increase I.Q. The researcher and NIH could then form lucrative licensing agreements with biotechnology companies for exclusive commercial exploitation of the genes. As such, Venter's patent claim has been aptly analogized to a "quick and dirty land grab"—equivalent to attempting to patent large tracts of land in the hope that some of the acres contain oil or gold.[10] Dr. James Watson, codiscoverer of the structure of DNA and former head of the $3 billion U.S. Human Genome Project, which is committed to mapping and sequencing each and every human gene, called the patent applications "sheer lunacy"; many believe his strong stance against the patents led to his resignation from the Human Genome Project in April 1992.[11]

Venter's patent applications, or similar applications by other government and private corporations, could create a unique and profoundly disturbing scenario. The entire human genome, the tens of thousands of genes that are our most intimate common heritage, would be owned by a handful of companies and governments. If Venter's applications or those of others are accepted, in a short time a few government bureaucracies and powerful corporations will have a monopoly on the use and sale of all human genes.

Not surprisingly, the patenting of human genes has led to massive legal and moral confusion, even among biotechnology's supporters. Kevin O'Connor, a senior analyst at the U.S. Office of Technology Assessment (OTA), confesses that few have even begun to think through all the arguments on whether the human genome or combinations of genes can be patented.[12] Others are concerned about the potential patenting of combinations of genes, including ultimately the entire human body. Derek Wood, head of the biotechnology patent office in London, comments:

This is clearly an area that is going to prove a pretty horrendous problem in the future. The difficulty is in deciding where to draw the line between [patenting] genetic material and human beings per se. Until we have a specific case in front of us, we have to make some pretty delicate judgments.[13]

Those "delicate judgments" will have to come sooner rather than later. According to published reports, the European Patent Office (EPO) has received patent applications that would cover women genetically engineered to produce valuable human proteins in their mammary glands. As discussed earlier, the genetic engineering of mammals (such as mice, cows, and sheep) to produce various valuable human proteins in their milk is ongoing and has shown some success. The EPO patent applications demonstrate that researchers want patent protection that would include any women who would be genetically altered in a similar manner. In one reported instance, Grenada Biosciences of Texas has applied to the EPO to patent genetically altered mammals created by Baylor College of Medicine in Houston and wants the patent to include both genetically altered animals and humans. Brian Lucas, a British patent attorney who represented Baylor College, has stated that the American patent attorneys who wrote the Grenada/Baylor application carefully crafted that application to include the coverage of women because "Someone, somewhere may decide that humans are patentable." The attempt to patent a "pharm-woman" who would produce valuable pharmaceutical products in her breasts has stirred considerable controversy in Europe. Paul Lamoye, the Belgian president of the European Green Party, called the reports of the patent applications "chilling news." He noted, "The fact that medical researchers and biotech companies have the audacity to even apply for it is clear evidence of the frightening direction this technology is taking."[14]

How could this be? How did our genes, the biological essence of humanity, become assignable commodities? How could living things or parts of living things, including the human body itself, be seen as patentable products indistinguishable from mechanical or chemical products? When was it determined that the building blocks of life belonged not to humanity, God, or nature, but to patent holders? And perhaps most perplexing, when and how was it determined that the historic question of the legal meaning of life was to be decided not by the people, but by the patent offices of the United States and Europe? The answers to these questions draw us back two decades, to the very definition of a "slippery slope."

Owning Microbes, Horses, and Honeybees

In 1971, Indian microbiologist Ananda Mohan Chakrabarty, an employee of General Electric (GE) in Schenectady, New York, set out to develop a special kind of "bug" that would eat crude oil. Its central use would be to devour oil slicks created by tanker spills or similar disasters. Nature had already produced several strains of bacteria that had the propensity to digest different types of hydrocarbons in oil. These bacteria were all from a family of bacteria known as *Pseudomonas*. Each of these bacteria had plasmids—auxiliary parcels of genes—that break up or "eat" oil. The trick was to somehow put them all together to get a "super" oil-eating bacteria. Chakrabarty performed this genetic manipulation feat by fusing together the genetic material from four types of *Pseudomonas*. Taking plasmids from three of the oil-eating bacteria, Chakrabarty transplanted them into the fourth, thereby creating a crossbred version with an enhanced appetite for oil. "I simply shuffled genes, changing bacteria that already existed," Chakrabarty explained. "It's like teaching your pet cat a few new tricks."[15]

In 1971, GE and Chakrabarty applied to the U.S. Patent and Trademark Office (PTO) for a patent on their genetically engineered microbe. After several years of review, the PTO rejected the Chakrabarty application on the basis that animate life-forms were not patentable. They noted that if either Jefferson or Congress had intended life to be patentable under the 1793 act, it would say so in the law. They also noted that in the few instances where life-forms had been patented—certain asexually reproducing plants—it was not the Patent Office that authorized patenting, but rather specially designed legislation passed by Congress.[16]

GE and Chakrabarty appealed the Patent Office rejection to the Court of Customs and Patent Appeals (CCPA). To the shock of many, they won. In a historic opinion, the CCPA, in a three-to-two decision, reversed the decision by the Patent Office and held that Chakrabarty could patent the oil-eating microbe. For the first time in patent law history, a court had allowed the patenting of a life-form. The opinion was direct: "The fact that microorganisms . . . are alive . . . [is] without legal significance." The legal distinction between living and nonliving matter had been rejected.[17] The court attempted to minimize its historic legal melding of animate with inanimate by noting that microorganisms were patentable because they are "more akin to inanimate chemical compositions such as reactants, reagents, and catalysts, than they are to horses and honeybees or raspberries and roses."[18]

The PTO was not impressed by the CCPA's legal decision to patent life, even lowly life-forms. It held steadfast to its original rejection of the GE patent and appealed the CCPA decision to the U.S. Supreme Court. The Supreme Court did not decide the case immediately. Instead, it instructed the CCPA to examine *Parker v. Flook,* a recent Supreme Court decision, which admonished courts that "we must proceed cautiously when we are asked to extend patent rights into areas wholly unforeseen by Congress."[19] The case warned courts that there should be no expansion of patent coverage without a "clear and certain signal from Congress."[20]

The Supreme Court's action indicated that it did not favor patenting life, and most felt that the patenting of life was almost certainly doomed unless Congress were to specifically enact legislation allowing it. But the CCPA was stubborn. In a three-to-two decision, it ignored the Supreme Court's advice. It reiterated its holding that microbial life was patentable. The Patent Office was equally convinced that life was not patentable, and once again appealed the CCPA decision to the Supreme Court. In October 1979, the Court decided to end the game of legal volleyball and announced that it would settle the question of the patentability of life, once and for all.[21]

Ironically, by the time the Supreme Court was to decide on the Chakrabarty patent, GE no longer had any intention of marketing its product. Apparently, while the microbe would eat oil in the lab, years of tests had proven it was too fragile to function in the open seas, where it was most needed. GE's motives in continuing its patent quest, despite the fact that it no longer had a viable product, were simple if not pure. It was using the Chakrabarty patent claim on the oil-eating microbe as an altruistic-seeming test case to establish the ground rules for the patenting of life. If the GE patent succeeded, the patent profit floodgates would be opened. GE and other corporations could then maximize their profits in the coming multibillion-dollar biotechnology industry.[22]

Given its extraordinary economic and ethical implications, it is surprising that the *Chakrabarty* case received so little public and media attention. Both the courts and the Patent Office seemed to view the case as merely a narrow legal issue in intellectual property or commercial law. Some, however, were more prescient. The People's Business Commission (PBC), run by Jeremy Rifkin and Ted Howard, filed an amicus (friend of the court) brief on behalf of the government's position on the nonpatentability of life. The PBC brief, written by Ted Howard, powerfully posited the importance of the *Chakrabarty* case:

> The case before the court may not appear to involve the life and death issues and passions of abortion, euthanasia or brain death rulings. Nonetheless, appearances aside, this case actually eclipses the import of these others because, in reaching a decision, a precedent-setting determination of the very nature of life will have to be decided upon. Whether such a definition is explicitly stated by the Court or not, hardly matters. If a ruling in favor of patenting genetically engineered living organisms is forthcoming, then manufactured life—high and low—will have been categorized as less than life, as nothing but common chemicals.[23]

The brief further warned that though the *Chakrabarty* case concerned a mere microbe, it could be used as a precedent for patenting decisions on all other life-forms including plants, animals, and perhaps eventually humans. And finally, the PBC asserted that with patent profits as fuel, the accelerated drive to commercialize engineered life would eliminate "all chance of meaningful public education and participation in the policy decisions surrounding genetic engineering, for the granting of patents is sure to escalate the drive toward commercial application. The genie will be out of the bottle before most Americans have even realized that the bottle was uncorked."

Some felt Howard's Cassandra-like warnings were unnecessary. The Supreme Court had clearly shown in the *Flook* case, two years before, by a six-to-three decision, that it would be extremely circumspect in expanding patent protection into unknown areas. The Court had already implicitly expressed reservations about this patent once, and it was sure to reject it—this time perhaps with a rebuke to the lower court. The Chakrabarty patent once again seemed doomed.

Oral argument before the Court was held on March 17, 1980. Three months later, the Court handed down its surprise opinion. The Supreme Court upheld the CCPA decision by a five-to-four margin. Chakrabarty was to be granted his patent. The highest court in the land had decided that life was patentable, stating that "the relevant distinction was not between living and inanimate things," but whether living products could be seen as "human-made inventions."[24] According to Chief Justice Warren Burger's majority opinion, the oil-eating microbe was not a product of nature, it was Chakrabarty's invention and therefore patentable. In coming to its precedent-shattering decision, the court seemed unaware that the inventor himself had characterized his "creation" of the microbe as simply "shuffling" genes, not creating life.

The four dissenters issued a terse and understated five-paragraph opinion written by Justice William Brennan. It concluded: "It is the role of Congress, not this court, to broaden or narrow the reach of patent laws. This is especially true where, as here, the composition sought to be patented uniquely implicates matters of public concern."[25]

All nine justices deciding the *Chakrabarty* case did agree on one thing. They specifically noted that this was a "narrow" case—one that did not affect the "future of scientific research."[26] Burger went out of his way to note that his decision in no way implicated "the gruesome parade of horribles," including the engineering of animals and man cited in the PBC brief.[27] The complete failure by the Court to correctly assess the impacts of the *Chakrabarty* decision may go down as among the biggest judicial miscalculations in the Court's long history.

It is reported that when the normally reserved Chakrabarty heard that the Supreme Court had ruled to allow patenting of his microbe, he startled his colleagues at the microbiology research center at the University of Illinois by shouting, "I won!"[28] This despite the fact that his genetically engineered "bug" was no longer marketable. Actually, the real winners emerged over the next few months on Wall Street as fledgling biotech companies like Genentech and Cetus, now buoyed by the prospect of patent profits, caused buying frenzies on the floor of the stock exchange. The 1980 initial public offering of stock by Genentech set a Wall Street record for the fastest price-per-share increase ever (from $35 to $89 in twenty minutes). This reportedly made "paper" fortunes of $50 million or more for Genentech's researcher-entrepreneurs like Herbert Boyer. The initial public offering of Cetus was nearly as spectacular, setting a record for the largest amount of money raised in an initial public offering ($115 million).[29]

The investment explosion did not surprise biotech analysts. From the beginning they did not share the Supreme Court's modest view of its opinion. Patent attorneys Arthur Gershman and Joseph Scafetta noted that the *Chakrabarty* decision took place on the centennial of the patent approval for Edison's incandescent lamp. They predicted that the biotech patent would have a more powerful impact on "the living standards in the 21st century . . . than the granting of the lightbulb to Edison in 1880 had on the 20th century."[30] Genentech declared, "The Court has assured the country's technology future."[31] Steve Turner of Bethesda Laboratories was equally enthusiastic: "It's an infusion of confidence that this country is serious about developing an important, domestic, homegrown, manufacturing, exporting industry."[32] Authors Sharon and Kathleen McAuliffe exclaimed, "Those working at the cutting edge of science and technology can rest assured that Jefferson's spirit lives on."[33]

Others continued to be concerned about the long-term consequences of the *Chakrabarty* decision. Ethicist and author Leon Kass summarized the view of many:

> What is the principled limit to this beginning extension of the domain of private ownership and dominion over living nature? Is it not clear, if life is continuum, there are no visible or clear limits, once we admit living species under the principle of ownership? The principle used in Chakrabarty says that there is nothing *in the nature of a being,* no, not even in the human patentor himself, that makes him immune to being patented. . . . To be sure, in

general it makes sense to allow people to own what they have made, because they artfully made it. But to respect art without respect for life is finally contradictory.[34]

The next decade was to prove both patenting proponents and opponents correct. Patenting did provide the trigger for a lucrative biotechnology industry. It also created the "gruesome parade of horribles" feared by Howard and Kass as the patenting of microbes led to the patenting of plants, and then animals, and finally human genes and tissues.

Of Men in Mice

Some called it the "mouse that roared," for others it was simply Patent No. 4,736,866, and for at least a few, it was an augury of the end of nature. On April 12, 1988, the U.S. Patent and Trademark Office (PTO) issued the first patent on a living animal. The patent was issued to Harvard Professor Philip Leder for the creation of a transgenic mouse containing a variety of genes found in other species, including chickens and humans. These foreign genes were engineered into the mouse's permanent germline in order to predispose it to developing cancer. The mouse was genetically engineered and designed to be a better research animal in which to test the virulence of various carcinogens.[35] While the media dubbed the patented animal the "Harvard mouse" because the patent was granted to Harvard researchers, this was a misnomer. The licensing rights for the patent are held by DuPont Company, the giant multinational that financed the Harvard research responsible for creating the genetically engineered mouse.[36]

DuPont got a lot more than just a genetically engineered mouse from the PTO. The patent licensed to DuPont is extraordinarily broad, embracing virtually any species of "transgenic nonhuman mammal all of whose germ cells and somatic cells contain a recombinant activate oncogene sequence introduced into said mammal, or an ancestor of said mammal, at an embryonic stage."[37] This means that DuPont has patent ownership of any animal species—be it mice, rats, cats, or chimpanzees—whose germlines are engineered to contain a variety of cancer-causing genes. The patent may well be among the broadest ever granted.

DuPont now markets the world's first patented animal, under the trademark name OncoMouse. The corporation has an attractive sales kit on OncoMouse that advertises the mouse's many advantages as a cancer "research model." According to DuPont, using Onco-Mouse will save the researcher time and money, in part due to the increased predictability of female mice in developing breast cancer.[38] The company justifies its patenting of OncoMouse by citing the importance of research into stopping the growing scourge of breast cancer. However, in actuality OncoMouse and other new and expensive research tools have done little or nothing to help in the war against breast cancer. An exhaustive 1991 federal study on breast cancer revealed that despite two decades of intensive and extremely expensive medical research, women's chances of surviving breast cancer have not substantially changed since 1970, and their chances for contracting the disease have risen alarmingly. "We must conclude that there has been no progress in preventing the disease," states Richard L. Linster, director of planning and reporting for the General Accounting Office (GAO), the congressional investigatory body that conducted the study. The GAO report's primary recommendation to help survival rates for the disease is more widespread use of mammography to detect early cancers.[39]

OncoMouse now has company at the patent office. On December 29, 1992, the PTO announced that it was awarding patents on three more genetically engineered research mice. The patents were granted to Harvard University (Leder's second animal patent), to GenPharm International of Mountain View, CA, and to Ohio University. Leder has said that he may call his second patented mouse, which has been engineered to develop an enlarged prostate gland, "Harvard II."[40]

Currently, well over 190 genetically engineered animals, including fish, cows, mice, and pigs, are figuratively standing in line to be patented by a variety of researchers and corporations.[41] It is only the complexity of the science, and the pressure caused by the political uproar around the patenting of animals, that has kept the PTO from further opening the animal patenting floodgates.

The PTO decision to patent genetically altered animals is a direct result of the misguided 1980 *Chakrabarty* decision by the Supreme Court. In 1985, the PTO ruled that the *Chakrabarty* holding, which allowed for the patenting of microbes, could be extended to include the patenting of genetically engineered plants, seeds, and plant tissue.[42] By 1987, the Reagan Administration's patent "slippery slope" had turned into a free-fall. On April 7, the PTO issued a ruling specifically extending the *Chakrabarty* ruling to include all "multicellular living organisms, including animals."[43] The radical new patenting policy suddenly transformed a decision about patenting microbes into one that allowed the patenting of all life-forms on earth, including animals. The ruling means that if a researcher implants a foreign gene or genes into an animal—for example, cancer genes into a mouse, or human growth genes into a pig—the genetically altered animal is considered a human invention under Jefferson's two-century-old definition of patentable "manufactures." The patented animal's legal status is no different than other manufactures, such as automobiles or tennis balls.

The revolutionary 1987 ruling on the patentability of animals, signed by Donald J. Quigg, Assistant Secretary and Commissioner of Patents and Trademarks, did have a silver lining for those concerned about the ethics of patenting life. The PTO ruling excluded human beings from patentability. The restriction on patenting human beings was based on the PTO's interpretation of the Thirteenth Amendment of the Constitution, the antislavery amendment, which prohibits ownership of a human being.[44] Unfortunately, there were several problems with the human exemption. For one, under the PTO's 1987 ruling, embryos and fetuses, human life-forms not presently covered under Thirteenth Amendment protection, are patentable, as are genetically engineered human tissues, cells, and genes. In fact, a genetically engineered "human" kidney, cornea, arm, or leg, or any other body part might well be patentable.

Neither the Reagan Administration nor patent commissioner Quigg could have been prepared for the controversy stirred up by their edict permitting animal patenting. Editorials across the country lambasted the new policy.[45] Bioethicist Robert Nelson reflected the general feeling about the decision when he stated, "It's a staggering decision. It removes one more barrier to the protection of human life.

Good God, once you start patenting life, is there no stopping it?"[46] In May 1987, a wide coalition of environmental, farm, and animal welfare groups, led by the Foundation on Economic Trends and The Humane Society of the United States, began pressing Congress to step in and stop the patenting of animals. In succeeding years, no less than nine bills were introduced in Congress to limit or prevent animal patenting. Though none made it into law, both the House and the Senate at different times passed legislation limiting patentability of life. Senator Mark Hatfield, Republican of Oregon, and a leader in the Congressional fight against animal patenting, summed up the arguments for many opposing the patenting decision on ethical grounds:

> The patenting of animals brings up the central ethical issue of reverence for life. Will future generations follow the ethic of this patent policy and view life as mere chemical manufacture and invention with no greater value or meaning than industrial products? Or will a reverence for life ethic prevail over the temptation to turn God-created life into reduced objects of commerce?[47]

The PTO decision also had far-reaching economic consequences. By providing the government's imprimatur on the genetic manipulation of animals and all other life-forms, it provided a federal guarantee that the raw material of the biotechnology age, all "multicellular creatures" including all animal species, could now be exclusively owned, used, and sold by corporate America. The new patent policy transformed the status of the biotic community from a common heritage of the earth to the private preserve of researchers and industry. Looking toward the future, the ruling set the stage for increasing competition among multinationals as they vie for ownership and control of the planet's gene pool, patenting everything that lives, breathes, and moves. In June 1992, at the Earth Summit in Rio de Janeiro, President George Bush faced international criticism as an "ecological villain" for his failure to sign an international agreement on species protection, biodiversity, and biotechnology. Bush rejected the treaty under pressure by U.S. biotechnology companies, who wanted to protect their abilities to patent the species of the world without restrictions.[48]

The most immediate economic impact of the patent policy could be felt in agriculture, where the major chemical, biotechnology, and pharmaceutical companies are positioning themselves to take over animal husbandry. By patenting genetically manipulated species, these corporate entrepreneurs could force farmers to pay royalty fees on every offspring created when the patented animal are bred, and every animal sold as part of their herds. America could see a new high-tech form of tenant farmers, indentured caretakers for corporate livestock species. For millennia people have owned livestock and other individual animals, and some have claimed that patenting of animals is only a logical extension of that ownership. However, human beings have never owned a species. No one has assumed the mantle of surrogate creator to own all giraffes, or all horses, and forced others to pay for the privilege of using these animals. As ethicist Leon Kass notes, "To own more of living nature than what one needs for one's own life and livelihood is hard to justify. Even harder to justify such monopoly when its sole purpose is to exclude others from similar benefits."[49]

The ultimate concerns about the patenting of animals are ethical. In a public statement issued in 1987, twenty-four religious leaders issued the following warning:

> The decision of the U.S. Patent Office to allow the patenting of genetically engineered animals presents fundamental dangers to humanity's relationship with the natural world. Reverence for all life created by God may be eroded by subtle economic pressures to view animal life as if it were an industrial product invented and manufactured by humans.[50]

A few months after the release of the religious leaders' statement on patenting, the *New York Times* singled it out for editorial criticism. In a lead editorial on February 22, 1988, entitled "Life Industrialized," the *Times* succinctly stated a view of life diametrically opposed to that of opponents of animal patenting, a view of life perfectly suited to the human body shop: "Life is special, and humans even more so, but biological machines are still machines that now can be altered, cloned, and patented."[51]

The Supreme Court's *Chakrabarty* decision has been extended, and continues to be extended, up the chain of life. The patenting of microbes has led inexorably to the patenting of plants, and then animals. Now the human body shop lobby is pushing toward the patenting of the human body, or at least the patenting of human cells, genes, and other biological materials. And just as the animal patenting debate has focused attention on the respect due to animals and our relationship to them as property, the patenting of human biological materials has produced a national legal tangle over who owns a person's genes and cells and exactly how far we can go in treating the body as patentable property.

15

Trafficking in human body parts raises the specter of thriving "used body parts" establishments emulating their automotive counterparts.

    Judge Ronald M. George, California Court of Appeals (dissenting in the Moore case)[1]

A Monopoly on Humanity

THE EARLIEST CONTROVERSIES about the patenting of human materials involved disputes over patenting "cell lines." It is ironic that the controversy about the ownership of the human body should be centered on the cell, the body's smallest component capable of carrying on all essential life processes. Yet cells have become extremely valuable. Scientists have now become so proficient in manipulating cells, by cloning and other regeneration processes, that they can grow and reproduce cells outside an organism.

A sample of cells that has grown through artificial laboratory cultivation is called a *cell line*. Cell lines have proven to be important as living tissues that researchers can use to study biologic processes and test the effect of chemicals. Unfortunately for researchers most cells do not grow outside an organism and cannot be used to create cell lines—for example, only one in 100,000 liver cells is capable of being grown in laboratories. However, not surprisingly, tumor or cancerous cells grow more easily than normal cells, and these cells have provided the basis for the most prolific cell lines used in research and experimentation.

Scientists developed the first human tumor cell line in 1951. The cells came from a virulent cervical cancer specimen removed from thirty-one-year-old Henrietta Lack.[2] These "HeLa" cells, named after the patient, grew quickly under artificial conditions. Too quickly. The HeLa cells not only grew, they proliferated out of control, invading adjacent cultures in many laboratories and accidentally contaminating numerous other cell lines. In order to avoid contamination problems, researchers eventually set up centralized "banks" of pure cell lines. This spared local laboratories from having to stock numerous cell lines and risk contamination. They could simply have the desired cell line sent to them as needed from the repository for a nominal fee (usually $40 or $50). The fee is low because the cell banks are nonprofit organizations, and the cells are donated by researchers from free patient materials. America's chief cell line repository is the American Type Culture Collection (ATCC) in Rockville, Maryland. It includes a variety of animal tissues, from the mountain goat to the pilot whale, and a variety of human tissue types. In 1989, the ATCC shipped 35,000 specimens to biomedical researchers around the world.[3]

During the last decade, patients have routinely signed consent forms allowing for disposal of cancerous or other "bad" tissue after surgery. Patients assumed that the tissue had no monetary value and that its only use was for nonprofit research. Now, however, biotechnology has made "bad" tissues a major body-parts business. Using new laboratory techniques, formerly "worthless" human tissue garnered through surgery can be manipulated to create human cell lines, which in turn excrete treasure troves of valuable biologic products such as hormones, anticancer agents, and antibodies. These new cell lines can create human biochemicals worth billions of dollars. The recent explosion in the value of human cell lines has caused a legal furor over cell ownership and the right to profits on cell lines, as researchers and corporations fight among themselves and with patients over the biological booty created by biotechnology.

Legal battles over human tissue ownership started almost as soon as researchers began mastering cell line technology. The first cell line developed from non-tumorous normal human cells was created in 1962 by a Stanford University microbiologist and resulted in a law case initiated by the scientist against the National Institutes of Health (NIH). The researcher was working under an NIH federal research grant when he created the historic strain of normal human cells in culture. After developing and cultivating the cell line, designated WI-38, the scientist formed a company to market the cells for use in the production of viral vaccines. The NIH claimed that the cells were federal property and charged the scientist with wrongfully exploiting federally funded research. The researcher resigned his position and filed suit, seeking title to the cells. The case was finally settled out of court in 1981, with the scientist retaining money from sales of the cells. The question of ownership of the cells was left unresolved.[4]

Meanwhile, in that same year, an even more convoluted case was brewing at the University of California at San Diego. Investigators at the university were developing a human cell line designed to secrete antibodies to fight cancer. A researcher, Dr. Heideaki Hagiwara, learned of the project and suggested to the university research team that they use the lymph cells from his mother, who was suffering from cervical cancer. The team agreed, and Hagiwara's mother's cells became the base of a cell line that, after development, successfully secreted valuable anti-tumor antibodies. Applying the precedent of the *Chakrabarty* decision, the cell line was patented by the university investigator. Subsequently, without the patent holder's knowledge, Hagiwara took a subculture of the cell line created through his mother's cells back to Japan, where he gave it to the Hagiwara Institute of Health (HIH), directed by his father. An initial agreement to use the cell line was worked out between the HIH and the patent holder, but that turned out not to be enough for HIH. The Hagiwaras went a historic step further and asserted ownership rights to the cell line and its product. They claimed that because the cells had originated with a member of their family, they owned the original tissue and were therefore entitled to a pecuniary interest in the cell line. In 1983, the parties reached an agreement under which the university retained all patent rights, and the Hagiwaras received an exclusive license to exploit the patent in Asia.[5]

A Lot Less for Moore

The early cases on human tissue ownership were only a warm-up to the highly publicized, multiyear legal battle waged by Alaskan businessman John Moore to retain a property interest in his own tissue. In the landmark case, perhaps the most famous of *Chakrabarty's* legal progeny, Moore sued the University of California, claiming that he was entitled to a share in the $3 billion of potential profits that the University and certain corporations could gain from a patented cell line developed from his cancerous tissues.

In the mid-1970s John Moore's life was in jeopardy, as he was diagnosed with an extremely rare form of cancer, hairy cell leukemia. Moore first visited the UCLA Medical Center on October 5, 1976, to confirm the diagnosis. After hospitalizing Moore and withdrawing extensive amounts of blood, bone marrow, and sundry other bodily substances, the attending physician, Dr. David W. Golde, gave Moore the bad news: He was indeed suffering from leukemia and he had "reason to fear for his life." In fact, the doctor reported, the disease had caused Moore's spleen to increase in weight from five hundred grams to nearly six kilograms. Golde strongly recommended that the cancerous spleen be removed. Moore signed a written consent form authorizing the splenectomy.[6]

Before the operation Dr. Golde and UCLA researcher Shirley W. Quan apparently suspected that the spleen tissue to be removed was capable of producing a remarkable blood protein—a protein that induces the growth of two types of valuable white blood cells that fight bacteria and cancer. Moore alleged that even before his splenectomy, Golde had made plans to keep a part of his spleen in order to investigate the possibility of utilizing these tissues for research and commercial purposes. On October 20, 1976, surgeons at UCLA Medical Center removed Moore's spleen. After the splenectomy a portion of Moore's spleen was taken to a separate research unit at UCLA, where it was stored and used for research. Part of this work included establishing a cell line from Moore's T lymphocytes (the white blood cells). This cell line, as expected, did produce valuable antibacterial and cancer-fighting pharmaceuticals.[7]

Meanwhile, Moore traveled from his home in Seattle to the UCLA facility several times between November 1976 and September 1983. He did so at the direction of Dr. Golde, based on the doctor's assertions that these visits were required for his health and well-being, and that his checkups had to be performed only at the UCLA facility by Dr. Golde. On each of these visits, Golde withdrew additional samples of blood, skin, bone marrow, and sperm. Moore now alleges that these visits were not done to further his medical treatment, but rather to help the researchers perfect his cell line for commercial use. He was never told by his doctors that they were developing a valuable cell line based on his cancerous tissue.[8]

It is undisputed that during the time Moore was under Golde's care, Golde and Quan were busy commercializing their new cell line. Among other activities, they advertised the cell line's unique capabilities to clients and provided various commercial firms with samples of the cell line and products created from it. On January 30, 1981, just a few months after the *Chakrabarty* decision allowing the patenting of life, the Regents of the University of California applied for a patent on the cell line. On the patent application, Golde and Quan were listed as inventors. With the lure of potential patent profits, Golde was able to enter into contracts in 1981, 1982, and 1983 with Genetic, Inc., Sandoz, Ltd., Sandoz United States, Inc., and Sandoz Pharmaceutical Corporation to collaborate on commercial exploitation of Moore's cell line. Genetics gave Golde 75,000 shares of stock at a nominal price and paid the Regents of the University of California and Golde $330,000 over three years. Sandoz paid Golde and the Regents $110,000. In March 1984, a patent was issued on the cell line that covers the cell line and nine products derived from it, including Immune Interferon, T cell growth factor, and a variety of other valuable human products. By established University of California policy, the Regents, Golde, and Quan share in any royalty profits arising out of the patent. Estimates of the cell line's ultimate worth have exceeded $3 billion. Though originally called the RLC cell line, purportedly to avoid detection by Moore, the cell line finally became known, and is now famous, as the Mo cell line.[9]

In 1984, Moore instituted his lawsuit to fight for his share of the profits created from his spleen tissue. His complaint stated no less than thirteen causes of action against Golde, Quan, the University, and the several biotech companies. Among other legal sins, the defendants were charged with lack of informed consent, conversion (an illegal taking of his property), fraud, deceit, and unjust enrichment. Two years later, after extensive legal maneuvers, Moore's case suffered a setback. The trial court dismissed Moore's complaint. The court flatly rejected his position that one could have a property claim on one's own discarded tissues.[10]

Moore appealed the lower court decision, continuing to insist that if others could make millions on his body parts, he should also have that right. In July 1988, Moore's persistence was rewarded. The California Court of Appeals reversed the lower court and held that Moore had a property right in his own bodily tissues and that he was entitled to a part-ownership in the patented cell line that bore his name. The court's opinion sent shock waves through the biotechnology industry, establishing for the first time that individuals owned a property interest in products developed from their cells. The daunting specter raised for the industry was a flood of claims for compensation by all those whose cells and tissues or genes were being used for living patentable products.

The appeals court was not ambivalent in its controversial holding: "The essence of a property interest—the ultimate right of control—therefore exists with regard to one's human body."[11] The court also held that Moore's consent to surgery did not imply consent for the commercial exploitation of his tissues.[12] The court was caustic in its rejection of the University's position that ethically no organ or tissue donor should have a property right in body parts, but that they (the University) could. In the words of Judge Rothman, writing for the majority,

> Defendant's position that plaintiff cannot own his own tissue, but that they [the University and the biotech companies] can, is fraught with irony. Apparently defendants see nothing abnormal in their exclusive control of plaintiff's excised spleen, nor in their patenting of a living organism derived therefrom. We cannot reconcile defendants' assertion of what appears to be their property interest in removed tissue and the resulting cell-line with their contention that the source of the material has no rights therein.[13]

In a spirited dissent, Judge George argued against the majority's opinion that the body was property that could be sold by its "owner." Among other concerns he was clearly alarmed at the public policy implications of opening a human body shop: "The absence of legislation regulating the trafficking in human body parts raises the specter of thriving 'used body parts' establishments emulating their automotive counterparts, but not subject to regulation comparable to that governing the latter trade."[14]

Golde, the University, and the other defendants appealed the case to the California Supreme Court. The decision, when it came in July 1990, was something of a compromise between the two prior court decisions. The California Supreme Court reversed the appellate court and held that human cells and tissue were not property like any other. Human tissues could not be sold or bartered by the person giving them up. Moore had no "property right" in the tissues of his body. However, Moore was not totally forsaken by the court. Rather, the court held that Dr. Golde violated a fiduciary duty he had to Moore by not fully informing him of the financial potential of his tissues: "We hold that a physician who is seeking a patient's consent for a medical procedure must, in order to satisfy his fiduciary duty, disclose personal interests unrelated to the patient's health, whether research or economic, that may affect his medical judgment."[15]

The court's decision means that Moore, while not able to claim a property right in the patent, will probably be able to claim some money damages for the breach of the University's fiduciary duty to him. The court's ruling in *Moore* caused the biotech industry to heave a collective sigh of relief. Their worst nightmare—thousands of tissue donors becoming part-owners in patented cell lines and other biotechnology patents and products—had been avoided.

For opponents of the commercialization and patenting of the human body, the *Moore* decision was a mixed blessing. The court had performed a valuable service by refusing to go along with the appeals court in breaking new ground in further reducing the body to commodity. A troubling body shop scenario that would have involved doctors, researchers, and patients bartering with one another for patent profits on tissues prior to medical operations had been avoided. Moreover, if the appeals court ruling had been upheld by the higher court, a dangerous precedent would have been set, undermining current bans on the sale of organs and fetal parts for transplantation. As noted by Judge George, the decision could have become a legal imprimatur for the establishment of organ, tissue, and gene human body shops not too dissimilar from auto body shops.

On the other hand, the opinion does nothing to alleviate the injustices and ironies spawned by the Supreme Court in the *Chakrabarty* decision. The court in *Moore* correctly denies the patient the right to sell his tissues, stating that human cells and tissues cannot be treated as mere commodities in the medical marketplace. Yet the court can do nothing to undo *Chakrabarty*'s grant of a government-granted patent monopoly on the sale and use of an individual's cells, tissues, and genes by patent holders. *Chakrabarty*, as modified by *Moore*, now means that Moore has no right of ownership of his body, but the University of California does. The organ or tissue donor must be governed by altruism, but the patent holders can make billions. In a dissent to the majority's decision in *Moore*, California Supreme Court Judge Broussard expresses the contradiction encoded by the holding in the case:

> . . . the majority's rejection of plaintiff's conversion cause of action does *not* mean that body parts may not be bought or sold for research or commercial purpose or that no private individual or entity may benefit economically from the fortuitous value of plaintiff's diseased cells. Far from elevating these biological materials above the marketplace, the majority's holding simply bars *plaintiff*, the source of the cells, from obtaining the benefit of the cells' value, but permits *defendants*, who allegedly obtained the cells from plaintiff by improper means, to retain and exploit the full economic value of their ill-gotten gains free of . . . liability.[16]

Invading the Body Commons

While Moore attempts to legally garner part of the profits gained from his cells, another patent uproar involving human cells has emerged. October 29, 1991, will go down as a historic day for the human body shop. On that day, for the first time, the patent office granted patent rights to a naturally occurring part of the human body. Patent No. 5,061,620 grants to Systemix, Inc., of Palo Alto, California, corporate control of human bone marrow *stem cells*, the progenitors of all types of cells in the blood. What makes the patent remarkable, and legally suspect, is that the patented cells were not any form of product or cell line. They had not been manipulated, engineered, or altered in any way. The PTO had never before allowed a patent on an unaltered part of the human body.[17]

Stem cells, which are used in bone marrow transplants and other blood treatments are difficult to isolate. Researchers at Systemix claim to have discovered a process that yields an unusually pure strain of stem cells. If proven, Systemix's process would represent a significant breakthrough in improving bone marrow transplants and in the development of new genetic therapies for leukemia and a variety of other blood disorders—even for AIDS patients. What stunned and outraged the scientific community was that the Systemix patent not only covers the process by which Systemix isolates human stem cells, it also covers the stem cells themselves. "It really is outlandish to believe you can patent a stem cell," asserted Peter Quesenberry, medical affairs vice chairman of the Leukemia Society of America. "Where do you draw the line? Can you patent a hand?"[18]

If the stem cell patent survives what are sure to be a series of court challenges, every individual or institution that wishes to use stem cells for a commercially viable cure for diseases or disorders will have to come to a licensing agreement with Systemix. Systemix now has its own privately controlled monopoly on human stem cells. As noted by ethicist Thomas Murray, "They've invaded the commons of the body and claimed a piece of it for themselves."[19]

The cell patent granted to Systemix, the thousands of brain genes whose patents are currently pending on behalf of Craig Venter and the NIH, and the almost two hundred animals lined up for patenting provide an important perspective from which to survey the influence of the Supreme Court's ill-advised decision in *Chakrabarty*. Most of the "gruesome parade of horribles" predicted by the those opposing the 1980 patent decision have become, with dizzying rapidity, realities. The distance traveled since the *Chakrabarty* decision in furthering the patenting of life can be highlighted by considering whether the court, in 1980, would have still ruled that life-forms were patentable had the organism in question not been a lowly, ineffective, oil-eating bacteria, but rather a human embryo, or a genetically engineered chimpanzee, or the entire set of over 100,000 human genes. The answer is clearly no. The Supreme Court would never have ruled for the patenting of these other living organisms or human subparts. Any such decision would have led to an immediate public uproar. Yet, through the slippery slope descent described in this and the preceding chapter, just such an expansion of patenting has occurred.

As government agencies refine the legal and technical means to patent numerous life-forms and untold thousands of humanity's genes and cells, we can only look with extreme trepidation at the future. There is little question that if it is not stopped, the patenting juggernaut will continue to transgress into life in all its forms. As research continues in cell analysis and in the deciphering of the human genome, corporations and researchers will fight for patent ownership of commercially valuable genes and cells. As there are advances in reprotech, human embryos may well be patented. Animals engineered to contain increasing numbers of human genes may soon be patented. Genetically engineered human body parts will almost certainly be patented. And looking into the more distant future, perhaps a genetically altered human body itself may be patentable. As patenting continues, the legal distinctions between life and machine, between life and commodity, will begin to vanish.

16

*A bokanovskified egg will bud, will proliferate, will divide.
From eight to ninety six buds, and every bud will grow into
a perfectly formed embryo. and every embryo into a full-
sized adult. Making ninety six human beings grow where
only one grew before. Progress. . . . ninety six identical
twins working ninety six identical machines!*
    Aldous Huxley, Brave New World

A Clone Just for You

WHEN IT FINALLY occurred—after years of science fiction fantasy—it was done so quietly and clandestinely that the world almost missed it. At a meeting of the American Fertility Society in Montreal in October 1993, Dr. Jerry Hall reported on his historic experiment. Hall's paper, written with his colleague Dr. Robert Stillman, was well received. It was awarded the "general program prize" as the best paper presented at the meeting. In his presentation Hall explained how he and Stillman had started with seventeen embryos and, using modest cloning technology, had multiplied them into forty-eight. Hall and Stillman had been the first researchers to clone human beings. At the meeting, the scientists suggested that the cloning of humans might be useful in the treatment of infertility.[1]

Soon after the Montreal meeting, Hall and Stillman were in the midst of a raging bioethical debate. While the scientists' modest cloning experiment may have seemed unremarkable to them, others recognized it as an epochal and perilous step for society. The

New York Times ran the front page headline "SCIENTIST CLONES HUMAN EMBRYOS, AND CREATES AN ETHICAL CHALLENGE." A spokesperson for the Japan Medical Association found the experiment "unthinkable"; French President François Mitterand pronounced himself "horrified"; the Vatican warned that the experiment could lead humanity into a "tunnel of madness." A poll taken by *Time* magazine and CNN showed that over seventy-five percent of Americans disapproved of cloning and thought it should be temporarily or permanently banned.[2]

Some were more positive. Many in the fertility industry proclaimed that they could now patent and sell "superior embryos" to couples who could not make embryos of their own. Clients could actually be shown pictures of good-looking, well-adjusted kids or adults and then be assured that they were receiving exact cloned embryonic copies of these individuals as future sources for organ and tissue transplants. Should a person become ill later in life, the cloned embryos could be gestated in paid surrogate mothers, and then aborted to obtain tissues or bone marrow for a perfect transplantation fit with the patient.[3]

While the majority saw the new reality of human cloning as a nightmare, many biotechnologists saw it as the culmination of decades of effort and research. For many years scientists had understood that natural reproduction was too slow and undependable as a method of "biofacturing" numerous identical copies of genetically engineered living organisms. A procedure was required that would mechanically reproduce exact duplicates of genetic material—endless "carbon copies" of engineered microbes, plants, and animals, including humans and human subparts. What, after all, is the use or profitability of isolating and patenting human and other valuable genetic material unless it can be copied and reproduced in industrial quantities? Why create and patent transgenic livestock and research animals unless they can be duplicated at will? And finally, why genetically engineer "superior" humans, or perfect body parts, if you cannot then make endless copies? From the outset of the genetic revolution, the challenge was clear: If we can manufacture millions of identical machines, books, clothing, or computers, why can't we biofacture millions of identical life-forms, including the human body and all its subparts?

Shocking Urchins and Duplicating Frogs

The history of cloning began over a century ago, as researchers began to manipulate a natural form of asexual reproduction called *parthenogenesis*. Parthenogenesis is a strange and unusual form of reproduction that occasionally occurs in plants and animals. The word is derived from the Greek *parthenos*, "virgin," to signify that in parthenogenesis there is no mixing of parental genes—all the genes come from one parent organism. It is a kind of "virgin birth." Parthenogenesis produces offspring by action within a single cell. Methods include cell division in bacteria, cell budding in yeasts, and vegetative duplication from plant cuttings.

Parthenogenesis has been observed in certain insects and animals. It has even been reported, in a few instances, in humans. One of the more plausible reports of human parthenogenesis occurred during World War II, in Hanover, Germany. In 1944 a woman, caught in an Allied bombing attack, collapsed on the sidewalk. Nine months later she gave birth to a daughter, who appeared to have identical fingerprints, blood type, and other indicators with her mother. The woman adamantly maintained that she had not had intercourse, and medical tests appeared to support her claim. How could the child have been conceived? Examining physicians hypothesized that the shock of the bombing attack may have somehow jarred a dormant body cell within the woman's uterus, triggering parthenogenic, nonsexual reproduction.[4]

Throughout the nineteenth century, scientists were fascinated with parthenogenesis. By the end of the century, pioneer experimenters with parthenogenesis were attempting to use chemicals on the eggs of various species to artificially induce parthenogenic creation of offspring. In 1896, German embryologist Oskar Hertwig added strychnine, or chloroform, to seawater containing sea urchin eggs; remarkably, this process fertilized the eggs without any contact with sperm. Three years later another German, Jacques Loeb, successfully repeated these sea urchin experiments. Over the next half-century, parthenogenic manipulation of sea urchin eggs increased rapidly. In fact, it seems to have gotten totally out of hand. By 1952, Dr. A. D. Peacock reported to a parthenogenesis meeting in Belfast, Ireland, that he had compiled over 370 ways of producing sea urchins without aid of sperm. These included 45 kinds of physical "jostlings," 93 chemical methods, 64 biological methods, and 169 combinations thereof.[5]

Most Americans became aware of the parthenogenic phenomenon when they saw an outwardly normal looking rabbit on the cover of a 1939 *Life* magazine. Despite its appearance, this rabbit was far from ordinary. It had been created by researcher Gregory Pincus as part of a group of experiments on human contraceptive techniques (ultimately, Pincus was to conduct pioneering work on the birth control pill). The unique offspring was created inadvertently when Pincus administered thermal shock treatment to a female rabbit. As with the sea urchins, the shock was sufficient to trigger the female rabbit's egg to fertilize. The one cell in the female rabbit was "shocked" into reproducing itself, and a baby rabbit was produced. No male was required.[6]

With the manipulation of parthenogenesis, humans took the first important step toward artificially reproducing life-forms. Significant advances in the understanding of human cells and genes were necessary before cloning would emerge as a viable technique, but there were tantalizing clues. As early as 1931, twenty years before the structure and nature of genes was first discovered, Dr. Ernest Messenger, in a book entitled *Evolution and Theology*, came up with the following revolutionary idea: "In principle, every cell in an ordinary organism contains . . . the virtuality of the species and the race." Based on this notion, seven years later Nobel Prize–winning zoologist Hans Apermann postulated a "somewhat fantastical" experiment. The idea was to remove the nucleus from an egg cell and replace it with the nucleus of some other cell to see if the transplanted cell nuclei would reproduce itself. As noted by Apermann, "This experiment might possibly show that even nuclei of differentiated cells can initiate normal development in the egg protoplasm."[7]

It took until 1952 for cloning theory to become practice. In that year Drs. Robert Briggs and Thomas J. King of the Institute for Cancer Research in Philadelphia, Pennsylvania, performed a landmark experiment replacing the nuclei of freshly fertilized frog eggs with foreign nuclei from another frog of the same species. Would the fertilized eggs result in offspring that were replicas of the frog who had donated the nuclei? The answer was yes—all the resulting tadpoles were duplicates of the tissue cell donor. Frogs had been cloned.[8]

Other successful experiments in creating frog clones followed, but genetic engineers would struggle for decades to achieve the far more difficult feat of cloning mammals. Nevertheless, the transition from human manipulation of parthenogenesis to cloning had been made. The basis for the mechanical reproduction of living organisms had been established. As summarized by author D. S. Halacy, "Here is the key point of cloning: the egg is essentially an environment; it is the implanted nucleus with its full set of chromosomes that determines the genetic makeup of a cloned individual."[9]

Xeroxed Cows and Freak Animals

The 60,000 visitors to the 1990 Dairy Expo in Madison, Wisconsin, had to have been surprised by the two-page ad in the Expo's unofficial program, a special issue of the magazine *Holstein World*. The text of the ad featured a top border with the repetitive come-on: "WIN A CLONE . . . AT THE DAIRY EXPO . . . WIN A CLONE . . . AT THE DAIRY EXPO . . . WIN A CLONE." The ad pictured an overlapping of five Xeroxed identical cows. The copy read in huge black letters, "A CLONE JUST FOR YOU." The ad explained:

> Genetic cloning is the newest, and one of the fastest ways to upgrade the genetics in your herd. To develop greater awareness of "cloning as a method of genetic advancement," Grenada Biosciences, Inc. and Holstein World are offering you an opportunity to win a "clone with outstanding genetics for milk and protein." . . . Remember come to Holstein World's booth during the World Dairy Expo . . . view the video on "cloning and other industry advancements" . . . and maybe there will be A CLONE, JUST FOR YOU![10]

Grenada's magazine advertisement at the Dairy Expo was a continuation of considerable publicity about the company's apparent success in cloning cattle. In February 1988, a front-page *New York Times* story announced Grenada's new-found ability to bring "factory-like efficiency to animal reproduction." The article described how the company had successfully used a modified version of the nuclear transfer technique utilized in the Briggs and King frog experiment to clone cattle embryos. Grenada's accomplishment was a milestone in the history of cloning—after three decades of difficult research, the widespread cloning of higher mammals was seen as possible and profitable.

The newspaper story described how Grenada's scientists had inserted the nuclei taken from a prize cow into the fertilized eggs of normal cows—eggs whose nucleic genetic material had been removed to make way for the new nuclei. Some of the altered fertilized eggs were transplanted into surrogate mother cows for gestation, while others were frozen for future research. The resulting calves were genetic replicas of the prize cow nucleus donor. According to the *Times,* the cloned calves now "romp alongside their surrogate mothers" at Grenada's Texas breeding farm.[11]

Grenada's work in cloning cattle capped a decade of progress in cloning a number of domesticated animals. Throughout the 1980s, researchers and corporations competed in a publicized genetics race to clone sheep and cattle. The two main contenders in this cloning competition were Danish scientist Sten Willadsen, the mastermind behind the Grenada clones, and Michigan-born Neal First, who through the funding of W. R. Grace was also able to clone cattle in his research facilities at the University of Wisconsin. The winner of the competition could reap enormous profits. If it can be perfected, cloning could significantly alter the $30-billion American beef industry and $18-billion dairy industry.

The procedure for cloning mammal embryos was developed by Willadsen as a result of research he conducted at the Agriculture and Food Research Council Institute of Animal Physiology in Cambridge, England. There Dr. Willadsen became most famous, or notorious, for his work in cell fusion, including the creation of the so-called geep—a bizarre animal resulting from the fusion of sheep and goat cells in embryos. After he left Cambridge, Willadsen spent a year at the Grenada Corporation before arriving in Canada at the University of Calgary and at Alta Genetics, Inc., also in Calgary. The procedure he pioneered was subsequently successful in cloning the embryos of rabbits and pigs as well as cattle.[12]

Willadsen's leading competitor, Dr. Neal First, is probably America's leading researcher in both the creation and cloning of transgenic cattle. Like Willadsen's, Dr. First's work also has resulted in breakthrough technology in cloning cattle embryos.[13] First's research has also been controversial. Wisconsin is dairy country, and First's research at the University of Wisconsin has been embroiled in the controversy over his attempts to create genetically engineered cows that produce more milk, cows that would overproduce milk in an already flooded market. As we have seen, hard-pressed dairy farmers are adamantly opposed to the use of genetically engineered bovine growth hormone (BGH) in cows to stimulate milk production. They are equally opposed to the concept, proposed by Dr. First, of creating thousands of cloned transgenic cows designed to excrete more milk; cows that would flood the milk market and lower milk prices for the farmer.

First's research grants have actually been threatened due to negative publicity about his research. First's sponsor, W. R. Grace & Co., is a massive multinational company specializing in chemicals and energy, which has had major image problems because of past negligence in handling a number of environmental pollution problems.[14] Grace, attempting to avoid more controversy, has been less than happy with First's knack for gaining publicity and his penchant for publishing his papers on genetic engineering and cloning without notifying his corporate benefactor. An October 1986 letter from Grace & Co. to the University of Wisconsin stated in no uncertain terms that funding for First's research could be jeopardized by "the potential for adverse publicity which could be associated with this [First's] research."[15] In a subsequent letter to First, the company stated that it was worried First's work might lead to a dairy farmer boycott of Grace's products. Grace also said that First's articles, including "Multiplication of Bovine Embryos" and "Nuclear Transplantation in Bovine Embryos," contained "language and key words that we believe may lead to unfavorable publicity." Additionally, the company warned First that his cloning article and others "contain information that may disclose patentable inventions according to our patent attorney." In its communication to him, the company asked First to remove the offending articles from publication.[16]

Grace's concern that First would jeopardize patent possibilities for the company was not hypothetical. In May 1991, it was announced that Grace's subsidiary, American Breeders Service

(ABS), had received the first patent ever granted for embryo cloning. The patent was for the process developed by First over the last decade. ABS is hoping that the patented cloning process will soon enable the company to make and sell many thousands of identical copies of the same dairy and beef cattle. Marv Pace, director of ABS Specialty Genetics, explains, "This technology removes the gamble in anticipating an offspring's sex, genetic makeup and production capabilities. Our goal is to have the first proven female clones available for market by the mid 1990s. We will then offer products including male and female high potential embryos to both dairy and beef producers." With an irony appreciated by those familiar with the Grace-First controversy over the researcher's sharing of scientific information, Peter Boer, executive vice-president of Grace, commented that the new patent is "indicative of the benefits of long term cooperative programs between industrial and university researchers."[17]

Grace was not alone in patenting the processes for cloning bovine embryos. Grenada had also secured several patents for cloning cattle. But in early 1992, just as the competition between the two companies seemed to be coming to a head, Grenada sold its operations and its patents to Grace.[18] What had prompted the reported downfall of the highly touted Grenada? A few months later, one possible answer was revealed. In March 1992, international media stories announced that Grenada's heralded cattle cloning procedures had gone badly awry. The headlines were arresting: "Lab Trial Spawns 'Freak' Animals." One story went on to tell of "freak giant calves born on farms in Britain and America." The stories described how Grenada's cloning process had gone "disastrously wrong." The problems centered on the cloned calves' weight. Instead of weighing the usual eighty pounds, several cloned calves swelled up to 150 pounds in the wombs of their surrogate cow mothers—emergency cesarean deliveries were required to deliver them. The unfortunate calves were created as part of a joint business venture between a British firm, Genus, and Grenada. As part of the business project, Grenada had recruited Sten Willadsen to develop cloned cow embryos for mass marketing. Using Willadsen's protocol, Grenada impregnated nearly a thousand cow "surrogate mothers" with the cloned embryos. But the project went under after it was found that one in five calves was emerging larger than normal, and that one in twenty was a giant. The freak calves have now been distributed to several American universities for study.[19]

The cloning "freaks" came as a shock to many. Kenneth Bandioli, currently president of the International Embryo Transfer Society, and formerly of Grenada, stated, "Clearly something about these embryos is not entirely normal."[20] Neal First called the super-large calves a "reasonable mystery. . . . No one has a good understanding of why the weight distribution is skewed."[21] Some took the failure in stride. Steve Amies, director for the British Milk Marketing Board's breeding production program, defended the cloning project: "We are more likely to get a uniform product with cloning. Any problems involved along the way, such as large calves, is [sic] part of the research. The scientists don't know what is going to happen until they do it."[22] Frank Barnes, president of Genmark, Inc., which has been offering cattle cloning services since 1991, insists that clone calves are "still an attractive product" to producers with top-of-the-line elite animals.[23] Others were shocked and repulsed by the news. They questioned the whole cloning enterprise. "The problems are a significant example of the difference between what the biotechnology industry promises and what it is able to deliver," noted John Stauber of the Foundation on Economic Trends.[24] British ethicist Dr. Andrew Linzey stated: "These are attempts to create animal machines that yield more milk, or tastier meat or bigger profits. To manipulate their lives for these ends is morally grotesque. When the end result is freak animals, inevitably there will be additional suffering."[25]

Lincoln Redux: Cloning Ourselves

As researchers struggle to resolve the problems that have suddenly sabotaged the cloning of livestock, and companies continue to battle over cloning patents and profits, society has now been confronted by the early stages of the cloning of humans. Researchers Hall and Stillman were aware that their pioneer experiment in human cloning would force the debate. "It was clear that it was just a matter of time until someone was going to do it, and we decided it would be better for us to do it in an open manner and get the ethical questions moving," Hall says.[26]

Hall's reason for assuming that some research group would soon clone humans if he didn't first was a technical advance that took place in 1991. As has been described, cloning animal embryos involves removing the nucleus from an embryo, copying it many times, and then substituting one of the copied nuclei for the original nucleus in any number of embryos. While this technique is practical for humans, it requires a large number of embryos to work. Currently, fertility centers simple do not have that many embryos to work with. However, Sandra Yee, Hall's colleague at the George Washington University School of Medicine, showed that it was possible to take human embryonic cells and give them an artificial coating that allowed each cell to grow as a separate embryo.

In their ground-breaking experiment, Hall and Stillman began with seventeen two- to eight-cell human embryos that had been fertilized at the George Washington IVF clinic. They then separated the cells from the embryos and coated them with Yee's artificial covering. The newly coated cells, now embryos, were placed in a nutrient solution where the embryos' cells began dividing again. The result: forty-eight new embryos, an average of three for each original embryo.[27] The final dream of Lederberg and his colleagues—the ultimate accomplishment of the human body shop—will have become a reality. Not everyone is sanguine about the cloning of humans. Dystopian visions of human cloning come from many biotechnology critics:

> Genetic engineering has the potential to create a vast army of identical clones, each produced to some preset specification. Cannon fodder, scientists, opera singers, all could be manufactured to order if the effort that went into putting men on the moon were directed to this new form of exploration. . . .[28]

Others, while admitting that human cloning is a scientific possibility, are less concerned:

> So we could engineer human giants tomorrow, and perhaps even clone them; but the prospect of genetically engineering mindless zombies, eager to do their master's bidding, doesn't worry me at

all. The existing methods of brain-washing and mind control already do that and more. . . . With simple technologies like those at your disposal, why bother with genetic engineering?[29]

Beyond the more futuristic debates, it is likely that the early stages of human cloning will imitate the more mundane commercial and utilitarian approach being taken with the engineering and replication of genes, plants, livestock, and research animals. Author Edward Yoxen has predicted that human cloning will follow the pattern of many of today's "gee whiz" medical breakthroughs, which more often than not help only the few and the wealthy. Yoxen envisions cloning pandering to a few individuals "with just enough cash to buy that kind of service."[30]

But who would such cloning customers be? What possible profit or utility is there to cloning human beings? Though we are still some distance from making cloning a commercial reality, scientists are already coming up with suggestions.

Embryo Research. Advances in genetics have made embryology a growing field. Researchers around the world are busy genetically screening pre-embryos, and doing research on fetuses and fetal tissue. The ability to create large numbers of identical embryos for research through cloning could provide scientists with several advantages. Cloning unlimited numbers of identical embryos could provide valuable predictability for researchers. Additionally, cloning, together with gene-splicing techniques, could offer researchers large numbers of identical, tailor-made, cloned embryos for various research projects. These research embryos could be genetically engineered to contain foreign genetic material or human genes predisposed to a disease or disorder being researched. Once cloned human embryos are created, it would not be surprising to see both the process of such embryo creation as well as the embryo "lines" themselves being patented by a sponsoring corporation. It is important to note that, as described in the last two chapters, current U.S. patent law makes patenting human embryos perfectly legal.

Organ Transplantation. Cloning could play an important part in "spare parts" surgery of the future. As we have already seen, the major problem with transplants is that the immune system of the recipient tends to recognize the transplant as foreign tissue and reject it. Genetics expert and author Jeremy Cherfas has suggested that cloning could provide one route to avoiding the rejection problem. Cherfas speculates that although it might be prohibitively expensive, a cloned "extra body" could be created for each individual. We could "remove its brain and keep it in suspended animation as a source of spare parts. A transplant from a donor of this sort would be like a graft between identical twins, with no fear of rejection."[31] If in the future brain transplants were made feasible, Cherfas's clones could be used to provide bodies for the brains of individuals whose original bodies may be destroyed by accident or disease.

Copying Ourselves. A growing number of single women and men are seeking nontraditional methods by which they can have genetically related children. Women have to undergo artificial insemination, men have to find a surrogate mother. Each current procedure involves the unwanted and unpredictable mixture in the offspring of another's genes. In the future women (and men, using nongenetic surrogate mothers) could use advanced cloning techniques to make "copies of themselves" without risking involving another's genetic input. Additionally, such cloning could be used to help certain infertile men and women obtain genetically related offspring—again using nongenetic surrogate mothers to carry their genetic clones.

Cloning Historical Figures. Two decades ago, Dr. Alof Axel Carlson of the University of California predicted that cloning would "permit the necrogeneticist to bring back individuals (e.g., historical personalities) of identical genotype to the dead by using sequences worked out from the entombed tissues."[32] At the time the idea appeared fanciful and extremely far-fetched. Recent events in "necrogenetics" seem to bring Carlson's bizarre vision a little closer to reality. In early 1991, the National Museum of Health and Medicine announced its intention to clone tissue samples of Abraham Lincoln. Dr. Victor McKusick of Johns Hopkins University requested permission from the museum to clone genetic material garnered from bone, hair, and bloodstains that the museum has preserved from Lincoln's autopsy. McKusick justified this unprecedented request on the grounds that the cloning might answer questions about Lincoln's health, including whether he suffered from Marfan's syndrome, a rare hereditary disorder.[33] Marfan's is a potentially fatal disease characterized in part by a tall, gangly appearance. Not surprisingly, a number of legal and ethical objections were raised to this proposed posthumous invasion of our sixteenth president's privacy. However, in May 1991 a nine-member panel of genetic experts sponsored by the American Medical Association gave a qualified green light to the proposal.[34] If final approval is given by the Armed Forces Institute of Pathology, which oversees the museum where Lincoln's remains are stored, researchers will attempt to create a library of Lincoln's genes. Critics have correctly cautioned that the Lincoln experiment could inspire widespread genetic cloning and examination of other historical figures. It remains to be seen whether future advances in cloning when applied to historical figures could fulfill Carlson's prophecy.

Eugenic Cloning. The use of human cloning that would most closely parallel its use in cattle would be in creating "prize" individuals. Many scientists have suggested that society would greatly profit from the cloning of geniuses and other exceptional people. Genetic authority Dr. Bernard Davis, of the Harvard Medical School, would like to see the cloning of those who excel "in fields such as mathematics or music where major achievements are restricted to a few especially gifted people. . . ." Author Dr. James Bonner claims that cloning could lift humankind toward "a new super species of human being." Of course, each visionary's dreams of desirable clones reflects his or her own particular point of view. Jeremy Rifkin notes that scientists interested in cloning rarely seem interested in cloning "social critics, reformers or revolutionaries."[35] Though the "positive" eugenic use of cloning may seem unlikely, it is not altogether unprecedented. As described in the chapter on artificial insemination, there are sperm banks that screen for superior traits, including one that specializes in the sperm of "superior" individuals—from Nobel Prize–winning scientists to champion athletes. Each year dozens of couples and individual women have sought this prize sperm in order to create super children. Would there be an equal if not greater demand for clones of such individuals?

———

As we begin cloning human beings, unprecedented legal and ethical questions arise. What of the products of early cloning experiments? If, as illustrated by Grenada's cloning of freak animals, fundamental errors in cloning techniques will need to be ironed out, how will we treat the ruined human clones who will result? What is the likely psychological impact of knowing that one is a clone? What are the eugenic implications of cloning—are clones a super race, or a subclass of humanoids without rights? The ultimate consequences of transforming human procreation into mechanical reproduction are unknown, but correctly feared.

PART FOUR

Closing
the Body Shop

17

He that troubleth his house shall inherit the wind.
Proverbs 11:29

*Those who cannot remember the past are condemned
to repeat it.*
George Santayana[1]

Crawling Machines
and the Invisible Hand

AS WE HAVE seen in our tour through the human
body shop, the market, aided by advances in medical and biological
technologies, has already seriously encroached on the body. We are
witnessing nothing less than a commercial invasion into our blood,
organs, and fetuses, our gametes and children, our genes and cells.
As body parts and materials are sold and patented, manipulated and
engineered, we also are seeing an unprecedented change in many of
our most basic social and legal definitions. Traditional understandings of life, birth, disease, death, mother, father, and person begin to
waver and then fall.

The current market onslaught places society at a critical juncture in its relationship with the human body. If we are to preserve the
body as inviolable, if childbearing is to remain inviolate, if we are to
retain our legal understandings of life, death, and the rights of parents and children, if we wish to protect the genetic integrity of our
own species and our fellow members of the biotic community: The

body shop cannot be allowed to continue business as usual. In fact, it must be closed.

But attempting to close down the body shop will not be possible without a better understanding of how the commercialization of human beings became acceptable. Even a detailed tour of the body shop does not fully explain how our society came to embrace the defective mythology of the body as commodity. For the commercialization of the body and its parts is not simply an unfortunate by-product of technology; nor is it merely the result of irresponsible science, poor regulation, greed, inaction, or governmental neglect. Rather, the body shop has its roots deep in our cultural, religious, and social history. In fact, today's cutting-edge commodification of the human body is the inevitable result of a unique way of thinking about nature, economy, and the human body introduced into Western civilization at the dawn of the modern age several centuries ago.

The history of the selling and engineering of life begins with the period traditionally called the Enlightenment. Thinkers of that era—Galileo, Newton, Kepler, Descartes, Locke, and others—changed forever the way we would look at the world and ourselves. They discovered new ways of examining nature, including the human body, and their work resulted in extraordinary advances in natural science and the development of technology. However, they also helped develop the view that life-forms are little more than sophisticated machines. This is known as the doctrine of *mechanism*. The human body was not spared this revolution in perception. One of mechanism's leading proponents, the noted French scientist Julien Offray de La Mettrie called humans "perpendicularly crawling machines."

The great thinkers of the seventeenth and eighteenth centuries also began to design a revolutionary new system for ordering human behavior. This new system was not based on traditional modes of social control—fear of God or church, duty to community or king. Rather, it was based on the principle that each individual would always act in his or her own self-interest. Francis Hutcheson, a seventeenth-century philosopher much influenced by Isaac Newton, called self-interest the "gravity" of social life. One of Hutcheson's students, Adam Smith, created the philosophic basis for what became known as the free-market system, with self-interest as the "invisible hand" that would lead to social justice and nearly endless wealth.

The dogmas of mechanism and the free market are the twin conceptual foundations of the human body shop. Mechanism is the basis for much of modern science. It has ingrained in our society the reductionist view of our bodies and other life-forms, which allows us to view them as biological technology available for sale. The ideology of the market has become the basic ordering principle of capitalist social life. It provides the primary rationale and ethical foundation for the selling of human body materials.

Even though only a small percentage of the current population has read the great Enlightenment thinkers, the doctrines of mechanism and the market have become second nature. In place of the religious awe of prior cultures toward the body, we commonly speak of our heart as a "ticker," our brain as a "computer," our thoughts as "feedback," and our digestive and sexual organs as "plumbing." It is not surprising that we then patent animals and body parts as "manufactures," develop a "new car" mentality about screening and selecting which of the unborn will survive, and attempt to make Xerox copies of living forms through cloning.

Moreover, "A contract is a contract," "The business of America is business," and "Let's get to the bottom line" have become platitudes by which we live, even when the issues are the contracting of motherhood, the business of blood, and the bottom line of marketing genetically engineered drugs to children facing discrimination.

Because the concepts of mechanism and the market have become intrinsic to our society, we cannot hope to halt the commodification of the body until we become more aware of these ideologies and how they came to govern so much of our lives and consciousness. Unless we first undertake an analysis of the historical and cultural influences that spawned the body shop, we cannot fashion effective biopolicies for the future that will protect us from the exploitation and commercialization of our bodies, limit the genetic manipulation of life, and preserve the integrity of our own and other species. We cannot change our future course without understanding the past.

In the next few chapters we will explore the historical and philosophical evolution of the body shop. We will give a detailed description of the growing influence of mechanism and the market on modern society. We are unfortunately unused to such analysis of current issues. Our seven-second sound bite culture discourages serious analysis of news events, and our mass media tenaciously resists placing current events in their historical or ideological contexts. As the English journalist G. K. Chesterton once quipped, "Journalism consists in saying, 'Lord Jones Dead' to people who never knew that Lord Jones was alive."[2] We hear urgent reports of wars, global environmental problems, technological breakthroughs, mass genocide, and other economic and social upheavals, without ever being given the chance to understand the deeper causes and meaning of such events. Scattershot coverage of current news isolates and makes meaningless the issues and stories reported. Each headline or news flash jolts the mind but then passes. No connection or comparison is made to related events in past days or years. The information never becomes cumulative or meaningful. Society tends to shake its collective head and move on.

So it has been with the human body shop. Disturbing headlines are routine. Over the last few years we have seen such alarming news flashes as:

TRADING FLESH AROUND THE GLOBE[3]
KIDNEYS FOR SALE[4]
NEW FETAL CELLS FOR OLD BRAINS[5]
MORE BABIES BEING BORN TO BE DONORS OF TISSUE[6]
FETAL TISSUES IN LAB MICE[7]
SUIT CHALLENGES CONSTITUTIONALITY OF SURROGATE MOTHERHOOD[8]
LAB TRIAL SPAWNS 'FREAK' ANIMALS[9]
AIDS VIRUS PLANTED IN MOUSE GENES[10]
U.S. SEEKS PATENT ON GENETIC CODE[11]
BREASTS PROVOKE PATENT CONFLICT[12]

Yet the overall picture of the human body shop is lost. Most people have little understanding of the philosophic and societal context of the commodification of the body. Few public voices articulate alternate values with which to treat our bodies, values antithetical to the body shop vision. As a result, the marketing and engineering of life continues unabated.

18

So God created man in his own image, in the image of God he created him; male and female he created them.
    Genesis 1:27

Living bodies are even in the smallest of their parts machines ad infinitum.
    Gottfried Wilhelm von Liebniz[1]

The Body as Technology

VIRTUALLY ALL PAST civilizations believed that the human form was sacred, made by God or the gods. In the Judeo-Christian tradition, it is understood that humankind was made in the "image" of God. For millennia this concept of the human and the human body, taken directly from the book of Genesis, formed the basis of Western ethics. Biblical understanding affirmed that all of life was divinely authored and definitionally "good." The Christian tradition gave the human body added significance through its belief that God became incarnate and suffered the joys and pains of being human. Subsequently, the body was given the ultimate sacramental value, as the early church taught that the union of human beings and God was to be obtained through ingesting the "body and blood" of Christ. And the body was to be exalted. The church promised those worthy that their bodies, albeit now "glorified," would be resurrected with them to heaven.

Theology often fails to translate into reality. Western history consistently chronicles the fact that despite its teachings on the sacred and numinous quality of the body, Christianity did not stop our inhumanity to our own species. The Christian centuries are replete with a variety of wars, persecutions, slavery, and other atrocities. Instance after instance affirms G. K. Chesterton's remark that "The Christian ideal has not been tried and found wanting. It has been found difficult and left untried."[2]

However, it is equally undeniable that the Judeo-Christian concept of humanity spawned unique advances in human rights and respect for the human person. The traditional teaching that the human form was divine in origin led to concepts key to modern justice. Inalienable rights, prohibitions on torture, church-mandated protection of civilians in war, acts of mercy, exalting of the poor and downtrodden, and respectful treatment of the dead are all descendants of Genesis. Whatever the vicissitudes of history, Western civilization shared a sacred communal image of our most intimate reality, the human body—an image that helped shape many of the truths the American forefathers held to be self-evident.

Today the traditional image of the human body has been shattered. Over the last centuries, and increasingly in recent years, our understanding and view of the body has undergone a conceptual free-fall as advances in science and technology appear to confuse and obscure any fixed definition of human life. Gradually, the body as sacred has evolved into the body as secular. The body is no longer seen as analogous to the divine, but rather as similar to the engines of industry. The image of God has given way to more modern deities. The body has become machine.

The mechanistic view of life is modern dogma. Recently, Dr. Robert Haynes, president of the 16th International Congress of Genetics, firmly reminded his audience that the doctrine of mechanism is a central organizing principle for the age of biotechnology:

> For three thousand years at least, a majority of people have considered that human beings were special, were magic. It's the Judeo-Christian view of man. What the ability to manipulate genes should indicate to people is the very deep extent to which *we are biological machines* [emphasis added]. The traditional view is built on the foundation that life is sacred. . . . Well, not

anymore. It's no longer possible to live by the idea that there is something special, unique, even sacred about living organisms.[3]

The scientist's proclamation was not an isolated view. A few months earlier, the *New York Times*, in a lead editorial entitled "Life Industrialized," bluntly argued that "Humans . . . [are] biological machines . . . that now can be altered, cloned and patented. The consequences will be profound but taken a step at a time, they can be managed."[4] Marvin Minsky, one of our nation's leading computer scientists, has called the brain a "meat machine," as he attempts to create a computer model for our minds.[5]

The dogma of mechanism gives us a defective mythology of our own bodies perfectly suited to the commercialization of body parts. In fact, we are so imbued with the doctrine of mechanism that we often perceive our body's organs, substances, subparts, genes, and cells as indistinguishable from the other mechanical and technological products in the marketplace. Mechanism has been further affirmed and accelerated by the discoveries and marketing schemes in biotechnology, reproductive technologies, and transplantation. Earlier in our body shop tour, we saw "defective" babies discarded with what one health care professional called a "new car" mentality. We witnessed courts legally defining the body as a "factory," and allowing for the paid "manufacture" of a baby by a surrogate mother. We followed revolutionary advances in the machinelike manufacturing of life through cloning. We observed governments and corporations patenting animals and human body parts just like any other "manufactures." We saw animals being created as "bioreactors" for valuable human genes, and the bodies of the dead used as "storage" for valuable organs.

Though it has blossomed in the age of biotechnology, the doctrine of mechanism is not some recent rubric dreamed up by overeager geneticists, surgeons, computer hackers, or editorial writers. Rather, it is a belief—virtually a substitute religion—that has been growing in the West for several centuries, since the beginning of the "age of machines." We cannot hope to counter the twentieth-century body shop until we understand this history of how we came to view the body as machine, how the body was transformed from sacred image to biological technology.

Galileo's Real Crime

One of the epochal moments in the history of Western science occurred on June 22, 1633, when Galileo, under extreme pressure by church inquisitors, "abjured" his heresy about the nature of the heavens. On that day he swore "to abandon completely the false opinion that the sun is the center of the world and does not move and the earth is not the center of the world and moves."[6] Since that time, Galileo has remained the ultimate symbol of modern enlightenment martyred by the forces of superstition and darkness.

Yet today Galileo is being charged anew. Many see the famous scientist as a key figure in the transformation of Western society's view of nature from *Terra Mater* (Mother Earth) to the mechanistic universe of modern science. Galileo's thinking is viewed as seminal to the seventeenth-century reduction of life to machine.

Galileo's revolutionary worldview can be partially understood through the fact that he was not a philosopher, but a mathematician. Galileo was convinced that the natural world could not be understood through metaphysics or spiritual studies, but only through quantitative measurement and rigorous mathematical analysis. The rest—memories, imagination, colors, tastes—were relegated by Galileo to the subjective and unmeasurable. In order to perceive the cosmos as mathematical and measurable, Galileo resurrected a controversial theory of matter from the Greeks and Romans. He believed that all things, animate and inanimate, could be seen as the interaction of tiny, ultimately measurable particles of matter. The doctrine was known as "atomism," and Galileo became one of its most renowned early proponents. Using this theory the great scientist maintained that the quantifiable and measurable were real, the qualitative and subjective unreal. Galileo felt that color and taste were opinions, while "atoms and the void [are] the truth." As noted by historian Lewis Mumford,

> Galileo committed a crime far graver than any the dignitaries of the Church accused him of; for his real crime was that of trading the totality of human experience, not merely the accumulated dogmas and doctrines of the Church, for that minute portion

which can be observed . . . and interpreted in terms of mass and motion.[7]

Mumford continues:

Galileo . . . surrendered man's historic birthright: man's memorable experience, in short, his accumulated culture. In dismissing [human] subjectivity Galileo had excommunicated history's central subject, multi-dimensional man. . . . Under the new scientific dispensation . . . all living forms must be brought into harmony with the mechanical world picture by being melted down, so to say, molded anew to conform to a more mechanical model. For the machine alone was the true incarnation of this new ideology. . . . To be redeemed from the organic, the autonomous, and the subjective, man must be turned into a machine. . . .[8]

Galileo was only one actor in the extraordinary revolution in human thinking that marked the seventeenth century. Philosopher Scott Buchanan calls the leading lights of the century—Bacon, Kepler, Galileo, Newton, Descartes—"world-splitters."[9] And they were. They separated out all nonmechanical aspects of nature and humankind and held them to be incapable of analysis and unknowable, and ultimately of little importance. They focused fully on treating nature in strictly mathematical and mechanical terms. In doing so, the creators of what they themselves called the "New Science" unquestionably set into motion the greatest revolution in thinking that has yet befallen our species. Each of these thinkers contributed essential elements to the transformation of our thinking about the natural world and our bodies. Each in similar ways assumed that nature worked analogously to a machine—specifically, the machine that they knew best, the clock. In this they took head-on the medieval view of the cosmos. Robert Boyle, the "father of modern chemistry," and among the first to encourage blood transfusions, wrote that the cosmos was essentially a clock, not an organic being as the medievals insisted. Boyle noted that the universe was

like a rare clock, such as that at Strasbourg, where all things are so skillfully contrived, that the engine being once set a-moving, all things proceed according to the Artificer's first design, and the motions . . . perform their functions upon particular occasions, by virtue of the general and primitive contrivance of the whole engine.[10]

The concept of God as clock maker and nature as clock created a new theological and scientific framework for the examination of nature. A succession of new discoveries about the geometric consistency of the solar system, the concept of gravity, physics, and the use of analytic geometry, led credence to the image of a clockwork universe. The explosion in the production of a wide range of efficient machines and automatons also gave support to the mechanistic theory of nature. And toward the end of the seventeenth century, Isaac Newton used the clockwork analogy as a powerful analytic tool for dealing with reality. His laws of motion and matter were seen by many of his peers as proof positive that the same "universal laws that governed the smallest portable watch also governed the movements of the earth, the sun, and the planets."[11]

It is easy to project too much of modernity on the great minds of the early enlightenment. The thinkers of that time had little inkling of the historic societal impacts that their new cosmology would create and were themselves somewhat disoriented in the transition that they had initiated. They were both revolutionaries and reactionaries. Each still had one foot in the medieval age and one in the modern, resulting in rather bizarre juxtapositions of beliefs. Johannes Kepler, who divined the geometry of the solar system, was equally fascinated by spiritual and angelic presences in the sun and planets. Isaac Newton, who "discovered" gravity and revolutionized optics, was also the premier alchemist of his day. René Descartes, who created analytic geometry and a new understanding of physiology, believed that he was simply clarifying the workings of the mind of God, a faith which, despite the claims of the church, was also that of Galileo.

Whatever the intentions of its authors, the mechanistic view of nature created by the Enlightenment is still our governing metaphor. As historian and philosopher William Barrett points out, "Our modern adventure began with the seventeenth century, but that earlier age has not vanished like a marker on a line that we have passed; it is still present, with all its paradoxes and tensions, in the uncertainties and malaise of our modern consciousness."[12] The influence of these thinkers was also key in the evolution of the human body shop. For the clockwork analogy was not only applied to the planets and other inanimate matter, it was also applied to the bodies of animals, including humans.

Beast Machines and the Death of the Soul

The playful puppy looking up at its master with mischief in its eyes, the stealth of a cat as it stalks an unsuspecting bird, the impatient circular motions of a lion caged in a zoo, the mournful cry of the dove in late afternoon, the innocence of a lamb just born—each of us has felt the delight, wonder, and melancholy that observation of and kinship with our fellow animals provides. Not so with the Enlightenment thinkers of the seventeenth and eighteenth centuries. For them, the physical and metaphysical status of animals created consternation and confusion.

As confirmed materialists and mechanists, many of the leading intellectuals of the scientific revolution stubbornly resisted any view of animals that would allow them emotions, intelligence, or soul. Their scientific belief led to continuous clashes with the church and others who felt that the scientific view was both contrary to common sense and common decency. The controversy over whether animals were ensouled beings or clockwork beast machines became a central philosophical battleground for over a century.

Francis Bacon was among the first to extend the clockwork analogy to the bodies of animals. In *Novum Organum* (1620) Bacon writes: "the making of clocks . . . is certainly a subtle and exact work: their wheels seem to imitate the celestial orbs, and their alternating with orderly motion, the pulse of animals."[13] The theories of Belgian anatomist Andreas Vesalius about human physiology, and the discovery by English physician William Harvey, in 1628, of the "pump"-like circulation of the blood, bolstered the growing seventeenth-century view that the body was machine.

However, Descartes is generally given the distinction of being the first to effectively translate the "New Science" into a mechanical analysis of living bodies. In his famous "Discourse on Method" (1637), Descartes argued that virtually all body movement and processes could be understood in terms of machinelike activity. Descartes maintained that the bodies of animals were "soulless automata" whose movements were little different than those of machines. This teaching was directly opposed to the teachings of the church. According to the traditional theological thinking of the time, each life-form had a "soul." Plants had a "vegetative" soul, which controlled their growth and form. Animals had a more advanced soul, the "sensitive soul," which gave them movement, intention, limited memory, imagination, and knowledge, but not individual immortality. Human beings were seen as having an immortal soul inseparable from the body. For Descartes, this whole theology was not in accord with observable fact:

> I wish you to consider . . . that all the functions which I have attributed to this Machine, such as digesting of meats, the beating of the heart and arteries, nourishment and growth, respiration, waking and sleeping; the reception of light, sounds, odors, tastes, warmth, and other similar qualities, into the exterior organs of sensation; the impression of the corresponding ideas upon a common sensorium and on the imagination . . . and finally the external motions of all the members of the body . . . I wish, I say, that you would consider all these functions as . . . neither more nor less than the movements of a clock or other automaton . . . so that it is not necessary, on their account, to conceive within it any vegetative or sensitive Soul. . . .[14]

As noted, the mechanistic concept of animals became a *cause célèbre*. For decades, the leading minds of Europe debated Descartes's *bête-machine* (beast-machine) theory. Traditional theologians attacked the Cartesian model as contrary to the teachings of the church. They also feared that the Cartesian view on animals would quickly extend to include the belief that humans were also without soul. Other church leaders were incredulous that Descartes and his ilk were attempting to reduce all of nature, even animals, to clocklike principles. Jesuit thinker Father Guillaume Bougeant wrote:

> I defy all the Cartesians in the World to persuade you that your Bitch is a mere machine. . . . Imagine to yourself a Man who should love his watch as we love a Dog, and caresses it because he should think himself dearly beloved by it, so as to think when it points to Twelve or One O'clock, it does knowingly and out of tenderness to him.[15]

Others, especially the Neoplatonist school of philosophers, also decried the fact that Descartes had taken the eternal "essences" —be they innocence, anger, dignity, or even wisdom—from the animal kingdom. They argued that these essences and emotions were clearly apparent to those making even the most superficial observation of animals. As philosopher Nicolas Hartsoeker remarked, " Of all Descartes's opinions none appears to be more extravagant or further removed from common sense than the one he has left us on the mechanical animal soul."[16]

Critics also noted that the Cartesian view of animals had deplorable impacts on animal suffering. In June 1640, Descartes was asked by a priest correspondent how it was that animals could feel pain if they had no souls. "They do not," responded Descartes, "for pain exists only through understanding of which brutes have none."[17] Given this view it is not surprising that the Cartesians became active adherents and practitioners of vivisection. There were many reports of Cartesians actively engaging in animal cruelty. Jean de La Fontaine's account of such experiments demonstrates the extent to which the Cartesian philosophy could lead to the desacralization of life.

> There was hardly a [Cartesian] who didn't talk of automata. . . . They administered beatings to dogs with perfect indifference, and made fun of those who pitied the creatures as if they had felt pain. They said that the animals were clocks; that the cries they emitted when struck, were only the noise of a little spring which had been touched, but that the whole body was without feeling. They nailed poor animals up on boards by their four paws to vivisect them to see the circulation of the blood which was a great subject of conversation.[18]

There were numerous condemnations of this view and the practice of vivisection. In 1648, theologian Henry More bluntly wrote Descartes that his theory was a "deadly and murderous" doctrine.[19]

Descartes's theory also attracted numerous followers. These adherents of rationalism increasingly saw the machine as a convincing analogue for biological processes. As advances were made in anatomy, they began comparing the body and its parts to a bewildering number of well-known mechanical devices. One of the greatest minds of the Enlightenment, Gottfried Wilhelm von Leibniz, maintained that "the machines of nature, that is, living bodies, are even in the smallest of their parts, machines *ad infinitum.*"[20] Another scientist saw the organs of the body "resembling pillars, props, crossbeams, fences, coverings; some like axes, wedges, levers, and pullies, others like cords, presses or bellows; and others again like sieves, strains, pipes, conduits and receivers."[21]

For some, however, Descartes had not gone far enough. For while Descartes relegated animals to machinelike status, he did not include humans under the same rubric. According to Descartes, human beings were unlike other animals. It was true they had a "machine" body, but they also had an immortal soul, a soul based firmly in the unique human power of reason. Other materialists saw no need for this dualism. For them, human life and behavior were essentially mechanical. With the help of the mechanistic view of consciousness advocated by seventeenth-century philosopher John Locke, even the mind was seen to obey mechanical principles. Humans were viewed as fully machines, both in mind and matter.

Among these materialistic critics of Descartes was Julien Offray de La Mettrie. La Mettrie was to take the final step in the philosophical mechanization of the human person. In his most famous work, *L'Homme Machine* (*Man a Machine,* 1748), La Mettrie posited the thesis that humans, just as other animals, were soulless machines. La Mettrie, while recognizing his debt to Descartes, differed with him in fundamental issues. Unlike Descartes, La Mettrie saw no need to distinguish the beast-machine from the man-machine. He saw no difficulty in espousing the view that mechanical laws produced both body motions and thoughts, and therefore also saw no need of a God to provide humans with souls. Here La Mettrie writes on both his debt to and differences with Descartes:

That famous philosopher (Descartes), it is true made many mistakes. . . . But after all, he understood animal nature. He was the first to have demonstrated perfectly that animals are pure machines. . . . All in all, whatever he may say about the distinction between man and animals, it is obvious that this is merely a trick and a writer's ruse, intended to make the theologians swallow a poison concealed behind an analogy that strikes everyone, and which they alone fail to see. For, it is just this strong analogy which compels all scientists and competent judges to confess that those proud and vain beings, men. . . . are, at bottom only animals and perpendicularly crawling machines.[22]

Though not always clear or consistent, La Mettrie's mechanistic view of the body was to have far-reaching influence. As historian David F. Channell notes,

By the end of the eighteenth century, mechanical philosophy seemed able to explain organic life from the function of bodily organs to the creative aspects of the mind. . . . In the reductionist world of the mechanical philosophy, machines and organisms could both be explained in terms of mechanical principles. The apparent conflict between the two is resolved by reducing life to technology. Life in general, even human beings, were at their base functioning as mechanical organisms.[23]

Ideas, of course, have consequences. The beast-machine concept and its subsequent extension to the human-machine provided a philosophical basis for the commercialization of life-forms and the human body shop. Historian Donald Worster captures the significance of the accomplishment of Descartes and his followers: "By reducing . . . animals to insensate matter, mere conglomerates of atomic particles devoid of internal purpose or intelligence, the naturalist was removing the remaining barriers to unrestrained economic exploitation."[24]

The beast-machine doctrine has contributed directly to the exponential increase in animal exploitation in modern times—from the assembly line destruction of animals in modern meat production, to the often cruel and unnecessary excesses of vivisection. The doctrine is epitomized in the current genetic engineering of the germ-lines of animals and their transformation into factories for producing valuable human biological products. The apotheosis of the beast-machine dogma is the new policy of patenting animals, which legally redefines the animal kingdom as "manufactures"—as machines.

La Mettrie's machine view of the human body, since upheld by generations of scientists and philosophers, allowed for the commodification of blood, organs, and even fetuses in the twentieth century human body shop. If we can sell "pulleys," "pumps," "pipes," and "conduits," why not their body counterparts? All that was lacking was the development of medical technologies that would give these parts value. As we have seen, advances in transplantation, reproductive technology and genetic engineering eventually provided the commercialization trigger.

———

The doctrine of mechanism was significantly refined as Western civilization entered into the industrial age. As more complex and sophisticated machines were developed, the machine image of the body also evolved. By the twentieth century, mechanism's proponents had set out to remake the body in the likeness of the modern motor with its greatest attribute—efficiency. These attempts were to have extraordinary consequences for human work and the human body shop. They were also to lead directly to perhaps the most pernicious practice of the twentieth century: eugenics.

19

*Man is a self-balancing, 28-jointed adapter-base biped,
an electro-chemical reduction plant, integral with the
segregated stowages of special energy extracts in storage
batteries, for subsequent actuation of thousands of
hydraulic and pneumatic pumps, with motors attached;
62,000 miles of capillaries, millions of warning signal,
railroad, and conveyor systems; crushers and cranes . . .
and a universally distributed telephone system needing
no service for 70 years if well managed; the whole,
extraordinarily complex mechanism guided with exquisite
precision from a turret in which are located telescopic and
microscopic self registering and recording range finders,
a spectroscope, [etc.]*
Buckminster Fuller[1]

*Those who mistake efficiency for meaning inevitably end by
loving compulsion.*
Owen Barfield[2]

The Human Motor

INDUSTRIAL SOCIETY BROUGHT a significant enhancement of the doctrine of mechanism. During the industrial age, scientists no longer viewed the natural world as the clockwork universe of Galileo or Newton, but rather as a great cosmic motor producing endless amounts of heat and energy. The body was no longer seen as an analogue to the relatively simple machines of Descartes's or La Mettrie, but as similar to the modern motor, modeled on a steam engine or electric power plant. As historian Anson Rabinbach notes, "During the nineteenth century, Descartes' animal machine was dramatically transformed by the advent of a modern motor, capable of transforming energy into various forms. . . . The human body and the industrial machine were both [seen as] motors that converted energy into mechanical work."[3]

The revolution in viewing the cosmos and the body as motors was girded by important discoveries in thermodynamics and the conservation of energy, especially the work of German physiologist and physicist Hermann Helmholtz. Helmholtz applied his theories to both machines and living organisms. As he noted, almost a century after La Mettrie, "The animal body therefore does not differ from the steam engine as regards the manner in which it obtains heat and force." He posited that "expenditure of energy can be understood in terms of an analogy between human labor and that of machines. The greater the exertion and the longer it lasts, the more the arm is tired and the more the store of its moving power is for the time being exhausted."[4]

Under the mandate of the new mechanistic vision, human and motor were wedded together both physically and psychically in the attempt to create ever greater production and wealth. As the industrial age advanced, the attempts to scientifically maximize production from both workers and machines led to a societal obsession with efficiency. By the late 1900s, "enlightened" policymakers, scientists, and captains of industry spoke endlessly of the necessity to use natural resources and human resources more efficiently. Philosopher Richard Weaver writes that for the modern industrial production enterprise, efficiency has become a "god" term—a proposition that has come to substitute for the religious and ethical codes of previous eras.[5]

There were significant problems for the new propagators of the "man as motor" myth. The human body was clearly not as "efficient" as the modern machine. In particular, they were confronted with the ever-present reality of human fatigue. As one scientist explained, "The muscle is an imperfect machine . . . a machine for work that is unilateral. . . . [T]he biceps can flex the forearm above the arm, and that is all; its action is exhausted by this work."[6] Fatigue and weariness, long seen as normal and even benevolent results of hard work, were now analyzed as a problem in the human machine. Even Karl Marx was concerned that the industrial system made fatigue inevitable. "Factory work exhausts the nervous system to the uttermost," Marx noted. "At the same time, it does away with the many sided play of the muscles and confiscates every atom of freedom, both in bodily and intellectual activity." To be successful in the industrial production process, Marx argued, the body will have to become more like the machine: "The main difficulty [of the factory system] is in training human beings to renounce their desultory habits of work, and to identify themselves with the unvarying regularity of the automaton."[7]

The human body had to be made more efficient, and the mechanical and social engineers took to the task with a vengeance. Just as they tested machines for efficiency and fatigue, the physiologists also conducted innumerable studies on human motion to find the most efficient ways to use the human machine. The clock, though no longer the dominant mechanical metaphor, was not lost in the world of the human motor. Efficiency meant minimum input for maximum output in minimum time. Every detail of human motion was carefully examined, drawn, and photographed in order to arrive at the perfectly efficient human motions. Psychologists contributed their part by conducting studies on mental fatigue and stress.

The effort to make the human body as efficient a production machine as possible reached its zenith with the work of U.S. mechanical engineer Frederick Winslow Taylor. In the years prior to World War I, Taylor introduced into the United States and Europe a system of "managing" workers for maximum efficiency. Taylor was a pioneer in "time and motion" studies of workers. He and his colleagues carefully examined every motion that factory workers made as part of the production process. Changes would then be recommended and carried out so that workers' motions would be optimally efficient. Taylor's revolution swept through the industrial nations over a period of several decades, directly affecting the production process for tens of millions of workers. Taylor's influence on modernity has been enormous. He furthered the ideal of efficiency and made it a reality for the general working population. As noted by economic historian Daniel Bell, "If any social upheaval can ever be attributed to one man, the logic of efficiency as a mode of life is due to Taylor."[8]

Helmholtz's human motor and Taylor's concept of efficiency are still principles under which many of us live. We feel the impact of these metaphors and ideas in our daily lives. As workers, we have for generations been viewed as little different from the machines we operate. Whether they are monitoring keystrokes or assessing workers on an assembly line, employers following in Taylor's tradition expect machinelike efficiency from workers. And despite the best efforts of Taylor and his followers, more often than not we cannot make the grade. We cannot endlessly repeat motions without becoming fatigued and ill. We are sensitive to noise and air pollution. Unlike motors, we need emotional security and support. We need meaning in our lives greater than mere efficiency or production.

Our modern workplaces, imbued with the "human motor" myth, end by destroying the body. Over 6.5 million Americans are hurt or become ill on the job each year. Almost half of those who become ill are victims of "repetitive traumas" to eyes, ears, and muscles—assaults on the body that are endemic to a mechanistic modern workplace designed to be antithetical to the body's natural movements. Computer work alone is causing an epidemic of disorders. A "human motor" who types sixty words a minute can make 18,000 keystrokes in an hour. This can lead to carpal tunnel syndrome and other serious nervous, muscle, and sight problems. Workplace injuries increase dramatically in periods of recession, as business demands even greater efficiency of output from fewer employees, and workers ignore symptoms of illness or fatigue for fear of losing their jobs.

In response to economic pressure and the continued influx of work-related diseases and injuries, the bioengineers and entrepreneurs of the body shop have begun extending the drive for more efficient workers in ways unimaginable to Taylor and the early pioneers of efficiency. Employers and their insurance companies are beginning to genetically screen and monitor workers for "undesirable" traits—genetic predispositions that would make them less efficient workers, or that would make them unusually susceptible to workplace-related disease or injury. As noted in a recent congressional report:

> Genetic information may be used by employers through genetic screening or genetic monitoring programs. . . . Individuals may be screened for genetic traits that render them susceptible to a pathological effect if exposed to specific agents present in the workplace. . . . Individuals also could be screened to detect general inherited characteristics not necessarily associated with occupational illness. Genetic information may be used to screen applicants for employment or current employees. . . . Once acquired

> the data may also be used to make decisions about the careers of
> employees. . . . Misuse of genetic information in employment,
> insurance and other areas is likely to involve some type of unfair-
> ness and discrimination.[9]

Numerous corporations have already begun the genetic screening or monitoring of their work forces. According to a 1990 report
by the congressional Office of Technology Assessment (OTA), 13 percent of Fortune 500 companies have used or are presently using some
form of genetic screening or monitoring on their workers. Employees will undoubtedly face even more screening in the future. According to the OTA, about 20 percent of companies think it is probable
that they will be using DNA screening of workers in the next five
years.[10]

Workers are also already being discriminated against due to their
genetic predispositions. Independent researchers, including Dr. Paul
Billings of the division of genetic medicine, California Pacific Medical
Center, have discovered numerous cases of genetic discrimination—
instances when employers have refused to hire an individual, and
where insurance companies have refused to cover individuals and
their families based on "poor" genetic readouts.[11] Discrimination
against genetically "inefficient" workers will undoubtedly increase
in the future, as more and more employers use genetic screening and
monitoring to limit their liability for workers who become ill on the
job and also attempt to assure themselves as genetically efficient a
work force as they can find. As genetic screening and monitoring is
increasingly used to discriminate against certain individuals in questions of employment and insurance, genetic privacy may become the
central civil rights issue of the next decade. Some legislatures have
already begun to take action. In 1992, Wisconsin became the first
state in the United States to forbid discrimination in employment or
insurance based on an individual's genetic readout.

The future use of biotechnology to create more efficient
"human motors" could go well beyond genetic screening. In the coming decades, society could witness the genetic engineering of many in
the work force. In the future, individuals could be pressured to undergo gene therapy to remove traits (for example, a predisposition to
alcoholism or depression) that might be viewed as undesirable by

employers and their insurers. In the more distant future, workers could be engineered to have greater resistance to workplace toxins or greater strength in physical areas required for a particular occupation. The new techniques for gene screening and somatic and germline gene engineering may give society the ultimate tools for achieving efficiency in production. Instead of changing our mechanistic workplaces to make them safer and more conducive to the human body, we can screen, monitor, or change the bodies of workers so that they better fit the modern workplace.

Our association of the body with "efficient machines" has crept into our culture in ways other than work. It has created a modern body type in the machine's image—what one commentator has called "techno-body."[12] The techno-body ideal, for men, and increasingly for women, is the "lean, mean machine": a hairless, overly muscled body, occasionally oiled, which very much resembles a machine. For many body zealots, the healthy body is one that functions and looks like an "efficient machine," not a body that is functioning in a natural and holistic fashion. Techno-body, of course, is not a natural form for the body, and many have become extremely dissatisfied with their body image. Polls indicate that over the last fifteen years, the number of men unhappy with their body image has doubled to include 34 percent of all men. Women dissatisfied with their overall appearance is even higher at 38 percent. Well over half of Americans now are unhappy with their weight or abdomen size, and significant percentages are deeply concerned about height, muscle tone, and facial characteristics.[13]

Millions go through daily body torture to achieve the techno-body ideal. Many pump iron (with the regularity of pistons working in a motor) to gain the requisite body hardness, a mechanical trait that is glorified. One commentator states that our society is going through "muscle madness." Machine manufacturers are selling $750 million worth of muscle-building equipment annually.[14]

At the same time, we eschew nonmechanical body softness at all costs, and weight remains our primary concern. In 1990, Americans spent approximately $33 billion on diets and diet-related services, much of it on "fad" diet systems that have a high recidivism rate. By the turn of the century, it is predicted we'll be spending $77 billion a year on losing weight.[15]

We are also attempting to achieve the new body image through surgery. The last decades have witnessed an astounding rise in cosmetic plastic surgery. It is estimated that each year around 1.5 million Americans undergo "aesthetic" surgery. The cosmetic surgery industry is reaping $4 billion a year—a figure many expect will double over the next decade. The fight against "softness" has also been a boon for the cosmetic surgeons. Over a quarter of a million Americans each year undergo liposuction surgery to remove fat. This represents a 250-fold increase over the last ten years. All in all, more is spent on fitness and cosmetics in the United States than on education or social services.[16]

In the future, cosmetic surgery could be augmented or even replaced by advances in the genetic engineering of the body. Currently, scientists are exploring ways to genetically engineer humans to increase size, decrease weight, increase muscle mass, and reverse aging. The use of human gene engineering for cosmetic purposes is termed "gene enhancement" therapy. Many in the biotechnology field believe that the future market for gene enhancement therapy could dwarf that of gene therapy used for curing disease.

The Super Race?

The complete triumph of efficiency over the frailties of our race necessitates something far more drastic than Taylor's enforced changes in workplace behavior, or even the current reshaping of the body with plastic surgery, obsessive muscle building, drugs, or gene enhancement therapy. The final solution to achieving a more efficient human body is more quality control in its reproduction, more control over the genetic heritage each human is to receive. If the body is to become closer to the machine, we will have to control how and with what materials it will be made.

Earlier in this book, we saw how the human body shop began this extraordinary process. The use of prize sperm and eggs in the new reprotech marketplace has started the drive toward predetermining the heredity of the newborn. Startling advances in the preimplantation genetic screening of embryos have initiated the journey toward the "perfect baby." With advances in somatic cell genetic engineering of humans, and discussions of germline manipulation of individuals, we have taken the first cautious steps toward actually controlling human evolution itself. The coming prospect of cloning humans, should it be possible, would offer the ultimate in quality control of human reproduction and genetic inheritance.

Though accelerated by the techniques of biotechnology, the attempt to control heredity in the interest of creating superior or more efficient humans predates the discovery of the structure of genes by several decades. Over a century ago, several scientists and social engineers began to formulate the principles and ethical bases for creating more efficient humans, and for preventing the birth of the unfit. They called the new science *eugenics*. The eugenic endeavor was steeped in mechanism and the ethic of efficiency. Eugenics, as described by one American leader of the movement, is "the science of the improvement of the human race by better breeding."[17]

The idea behind eugenics was to use techniques of selective breeding, originally used for plants and animals, on humans. As noted in the early 1900s by Charles R. Van Hise, president of the University of Wisconsin: "We know enough about agriculture so that the agricultural production of the country could be doubled if the knowledge were applied . . . we know enough about eugenics so that if the knowledge were applied, the defective classes would disappear within a generation."[18] Noted eugenicist and utopian socialist John Humphrey Noyes stated, "Every race horse, every straight backed bull, every premium pig, tells us what we can do and what we must do for man."[19] A more recent commentator, author Julian Huxley, argued,

> It is clear that for any major advance in national and international efficiency we can not depend on haphazard tinkering with social and political symptoms or ad hoc patching up of the world's political machinery, or even on improving education, but must rely increasingly on raising the genetic level of man's intellectual and practical abilities.[20]

The term "eugenics," taken from the Greek root meaning "noble or good in birth," was coined in 1883 by Francis Galton, who is generally regarded as the father of the eugenics movement. Born in 1822, the same year as Gregor Mendel, he was a cousin of Charles Darwin. Galton was inspired by his relative's epochal work, *Origin of Species by Means of Natural Selection on the Preservation of Favored Species of Life.* (It is interesting to note that today 80 percent of the title of Darwin's classic is dropped; perhaps the eugenic implications are partially to blame.) Darwin himself believed in favored and unfavored races of humans. He states:

> I could show [that war had] done and [is] doing much . . . for the progress of civilization. . . . The more civilized so-called Caucasian races have beaten the Turkish hollow in the struggle for existence. Looking to the world at a not very distant date . . . an endless number of lower races will have been eliminated by the higher civilized races throughout the world. [21]

Hoping to fulfill his famous cousin's prophecies, Galton devoted himself to the task of ensuring that "the more suitable races or strains of blood [are given] a better chance of prevailing speedily over the less suitable."[22] Galton summed up his hopes for humankind by noting that just as it was possible "to obtain by careful selection a permanent breed of dogs or horses gifted with peculiar powers . . . so it would be quite practical to produce a highly gifted race of men" by similar means.[23] Galton was an eccentric man, and the reasons for his passionate devotion to the cause of eugenics remain unclear. Whatever the personal bases for his eugenic beliefs, Galton, an atheist, found the search for making more efficient humans a surrogate religion. "An enthusiasm to improve the race is so noble in its aim," he declared, "that it might well give rise to the sense of a religious obligation."[24] As noted by historian Daniel J. Kevles, "[Galton] found in eugenics a scientific substitute for church orthodoxies, a secular faith."[25] Galton's followers showed the same quasi-religious zeal as their mentor. George Bernard Shaw stated: "There is now no reasonable excuse for refusing to face the fact that nothing but a eugenics religion can save our civilization from the fate that has overtaken all previous civilizations."[26]

The "science" of eugenics has two tools, negative eugenics and positive eugenics. The purpose of negative eugenics is to prevent the unfit, the "cacogenic"—historically, those viewed as insane, feebleminded, criminal, or in any other way inferior—from propagating. The function of positive eugenics was to encourage the propagation of the "aristogenic," those with what were believed to be superior traits.

The eugenics movement can be roughly divided into four stages. The first, from about 1870 to 1905, was a time of preparation, as the ideas of breeding humans in a more efficient manner became widespread. Galton and his followers were intent on proving that biology was destiny. They often opposed laws for the amelioration of the conditions of the poor on the grounds that such laws only encouraged the propagation of the unfit. Additionally, the early eugenicists supported restrictions on marriage and sexual behavior of the unfit. In 1896, Connecticut became the first state to regulate marriage for breeding purposes. The law provided that "no man and woman either of whom is epileptic, or imbecile, or feeble minded" shall marry or have extramarital relations "when the woman is under forty-five years of age." The statute set a minimum penalty of three years' imprisonment for violations. Within the next decade, five states had followed Connecticut's lead and passed eugenic marriage laws.[27]

From 1905 to about 1940, eugenics entered its second and most influential phase. Over a nearly four-decade period, eugenics was to become official state policy for countries around the world. During this period, the United States became the unquestioned leader of the eugenics movement. By 1910, a eugenics center in Cold Spring Harbor, New York, had been set up as a result of a $10 million grant by the Carnegie Institute. Soon eugenic societies sprang up in cities across the country. Famous scientists were exclaiming that "To go to the root of [America's] troubles, a better breed of men must be produced, one that shall not contain inferior types. When a better breed has taken over the business of the world, laws, customs, education, material conditions will take care of themselves."[28]

Eugenics at this time also contained a heavy dose of class politics. Many of eugenics' most ardent advocates were also members of America's ruling families. The country's bluebloods—the Harrimans, Lodges, Saltonstalls—were becoming increasingly concerned over their potential loss of political control to various immigrant groups.

After the immigration waves at the end of the nineteenth century, the "WASP" hegemony was being challenged by the Irish, the Jews, the Italians, and other newly arrived groups, who were beginning to elect their own leaders to power. Eugenics and the eugenic ideal was one weapon in the arsenal of the old elite to attempt to stem the immigration tide and undermine the new power base. They had some success. Eugenics was effectively used throughout this period to enact federal laws restricting immigration.

The main method for the implementation of negative eugenics in the United States was the sterilization of the "unfit." Sterilization for those incarcerated in mental institutions and in prisons became routine. The first law allowing for sterilization for the unfit was passed in Indiana in 1907, and fifteen other states followed suit in the next ten years. By 1930, twenty-eight states had passed sterilization laws, and 15,000 men and women had been sterilized under them. By 1939, nearly 30,000 had been sterilized. By 1958, almost 61,000 "unfit" Americans had been involuntarily sterilized under thirty state laws.[29] The majority of these statutes were based on the Model Eugenical Sterilization Law, drafted by Harry H. Laughlin, superintendent of the Eugenics Record Office. Laughlin's law called for sterilization—"regardless of etiology or prognosis"—of criminals, mental patients, the feebleminded, inebriates, the blind, the diseased, the deaf, the deformed, and the dependent (that is, the homeless—orphans, tramps, and paupers).[30]

Sterilization and negative eugenics were not unopposed. Powerful religious institutions strenuously resisted state interference with the rights of the "unfit" to marry or bear children. A Catholic theologian of the time wrote of the "unfit": "[The church] counts the earthly existence of a helpless cripple, a chronic invalid, or a mental weakling intrinsically good, and she knows that all such persons are capable of a life of eternal happiness face to face with God."[31]

As state-sponsored sterilizations increased and moral arguments reached a fevered pitch, the U.S. Supreme Court accepted a case that would resolve legal doubts about the sterilization of the "unfit." The case involved the proposed sterilization of a Virginia woman, Carrie Buck. Carrie, believed to be the daughter of a "feebleminded" mother, had been adopted at the age of four by a family in

Charlottesville, Virginia. There she attended school until the sixth grade. However, Carrie, who was believed to be mentally impaired herself, was unable to continue her schooling any further. After she left school, Carrie did housework under supervision. She proved "unmanageable" and subsequently became pregnant. In January 1924, Carrie, like her mother before her, was institutionalized at the Virginia State Colony for Epileptics and Feebleminded. Her child was born at the institution.

Several months later Carrie was chosen to be the first victim of the newly passed Virginia sterilization law. The law was challenged and began making its way through the courts. Genetics experts testified that Carrie had the mental age of a nine-year-old and that her daughter was also defective. They insisted that her feeblemindedness as well as her "immorality" were inherited and passed on. The state also had a practical argument. It maintained that if Carrie was sterilized, she could be deinstitutionalized and that significant expenditures to the state would be saved. The order for sterilization was upheld by both the circuit court and the Virginia Supreme Court. In the spring of 1927, Carrie's challenge to the Virginia law reached the United States Supreme Court.

The case, *Buck v. Bell*, became famous, or infamous, as did the written opinion by Justice Oliver Wendell Holmes. The Court upheld the law and the sterilization of Carrie. Holmes was himself an avid supporter of eugenics. His opinion for the court majority was an enthusiastic tract for sterilization and negative eugenics. His opinion contained the oft-quoted passage: "It is better for all the world, if instead of waiting for their imbecility, society can prevent those who are manifestly unfit from continuing their kind. . . . Three generations of imbeciles are enough." [32]

The American model of sterilization proved influential. Five years after *Buck v. Bell*, Denmark, Finland, Sweden, Norway, and two provinces in Canada had all initiated forced sterilization. In 1933, Hitler put into effect his notorious sterilization program. Rudolph Hess issued his famous statement, "Nazism is applied biology." [33] The Nazis directly credited the U.S. eugenics program as the basis for their own. One Nazi author approvingly cited the United States as a country where "racial policy and thinking have become much more popular than in other countries."[34] Historian Robert N. Proctor notes that "German racial hygienists looked to the United States for inspiration; in this sense, German racial hygiene followed the American lead."[35] Oddly, England, the birthplace of the eugenic movement, never was able to pass sterilization laws.

By the 1930s, eugenics critics were becoming more vocal. In 1934, a special blue ribbon committee of the British government, headed by Laurence G. Brock, issued a report seriously undercutting the scientific basis for eugenics. The Brock report noted that "the more closely individual records are examined the more difficult it becomes . . . to say with certainty that the genetic endowment of any individual is such that it must produce a given result."[36] In 1936, a committee of the American Neurological Association, headed by the Boston psychiatrist Abraham Myerson, came to a similar conclusion. The Myerson report, appearing less than a decade after *Buck v. Bell,* flatly contradicted the Supreme Court's logic in upholding sterilization: "There is at present no sound scientific basis for sterilization on account of immorality or character defect. Human conduct and character are matters too complex a nature, too interwoven with social conditions . . . to permit any definite conclusions to be drawn concerning the part which heredity plays in their genesis."[37]

A few years later, a significant legal blow was dealt to U.S. sterilization laws. Though lower courts continued to uphold sterilization statutes under *Buck v. Bell,* a 1942 Supreme Court decision, *Skinner v. State of Oklahoma,* began to punch holes in the constitutionality of such laws. The case involved the sterilization of convicted felon Jack T. Skinner. Skinner had been convicted twice for chicken-stealing, and a third time for armed robbery. He was to be sterilized under a law passed in Oklahoma in 1935. The law authorized sterilization for anyone convicted of a felony three times. However, it allowed exemptions for several "white collar" felonies such as convictions on embezzlement, liquor law violations, or political offenses. The Supreme Court unanimously struck down the Oklahoma law as unconstitutional. The opinion by Justice William O. Douglas resonates with the horror of Americans just becoming aware of the eugenic nightmare ongoing in Germany. It decries the Oklahoma law's discriminatory distinction between various types of felons. As held by the court: "When the law lays an unequal hand on those who have committed intrinsically the same quality offense and sterilizes one and not the other, it has made as invidious a discrimination as if it had selected a particular race or nationality for oppressive treatment."[38]

After World War II, eugenics lost even more of its allure. The savage example of the German extermination program of Jews, gypsies, and other "undesirables" put what seemed to be a permanent damper on the eugenics movement. The "science of eugenics," with its simplistic association of facial features with inferiority, unsophisticated view of genetic inheritance, and use of bizarre pseudosciences like phrenology, seemed totally discredited. Citizens and scientists alike were disgusted by the application of eugenics to exclude immigrants, to justify sterilization of criminals and the mentally ill, to prohibit marriages between different racial groups, and to exterminate those viewed as genetically inferior. The social scientists of the 1950s and early 1960s deemphasized genetics as a source of dysfunctional human behavior and focused on environmental factors as being the key to reforming those viewed as socially undesirable.

The New Eugenics

In the 1960s, eugenics began to resurrect. Prominent scientists once again began to see eugenics as the main hope for humankind's survival and prosperity. The fourth and current stage of eugenics was based on a revitalization of the science of genetics, especially knowledge about genetic abnormalities and their relationship to hereditary disease. A notable example was Jerome Lejeune's discovery in 1959 of the chromosomal abnormalities responsible for Down's syndrome. Identification of the genes responsible for sickle-cell anemia, thalassemia, and other single-gene hereditary disorders followed in subsequent years. As advances in understanding genes followed one another over the next twenty-five years, hundreds of genetic "defects" that caused hereditary diseases or disorders were located.

A new eugenics revolution had begun. This time the eugenic revolution was based not on political or racial prejudice, but rather on the "hard" science of genetic predisposition to physical or mental disorders or weaknesses. And the tools of the new eugenics were not clumsy and monstrous sterilization or extermination programs, but rather sophisticated techniques in genetic screening, embryo manipulation, and genetic surgery.

The new eugenics movement still relies primarily on negative eugenics. There is currently no cure for the vast majority of genetic diseases and disorders that have been identified. The sad fact is that the major "benefit" derived from the explosion of new information on the genetic predispositions of individuals is women's use of selective abortion to prevent the birth of "bad" gene carriers. In the future, genetic engineers may be able to use germline genetic surgery to alter an individual's genetic makeup to ensure that undesirable genes are not passed on to future generations.

Whether the tools are abortion or germline genetic alterations, the new eugenics movement, almost certainly, will seek to eliminate numerous "bad" genes that have little to do with serious physical disease. Scientists are currently attempting to find genetic links to a wide group of complex emotional and behavioral problems. With these new research efforts comes a troubling resurgence of biological determinism, the concept that biological and not environmental factors are primarily responsible for human behavior. Over the last few years, front-page stories have appeared in the United States and around the world heralding the genetic connection to manic-depressive disorder, schizophrenia, panic attacks, alcoholism, and even shyness. These genes and others linked to various undesirable behavioral traits will undoubtedly be targeted for elimination by the new eugenicists.

Even the modern-day Jack T. Skinners may not be safe. While sterilization of criminals is not in fashion, drug therapy and genetic engineering may be in the offing. Even our federal government may be abandoning the view that criminal behavior can be cured through education or rehabilitation, and reverting back to a genetic theory to account for crime. A conference that was to be partially sponsored by the National Institutes of Health (NIH) in 1992 on "The Genetic Factors in Crime: Findings, Uses, and Implications" contains the following, apparently official, statement on crime and criminals in its brochure:

Researchers have already begun to study the genetic regulation of violent and impulsive behavior and to search for genetic markers associated with criminal conduct. Their work is motivated in part by the early successes . . . on the genetics of behavioral and psychiatric conditions like alcoholism and schizophrenia. But genetic research also gains impetus from the apparent failure of environmental approaches to crime—deterrence, diversion and rehabilitation—to affect the dramatic increases in crime, especially violent crime that this country has experienced over the last 30 years. *Genetic research holds out the prospect of identifying individuals who may be predisposed to certain kinds of criminal conduct . . . and of treating some predispositions with drugs and unintrusive therapies* [emphasis added].

Only days before the conference was to be held, the NIH was forced to withdraw its support due to congressional and public outrage. Lacking federal funding, the conference had to be canceled.[39]

Despite public controversy, the new findings of genetic links to what was formerly thought to be socially controlled behavior are being highly touted. A gullible press and the growing willingness among scientists to project all our social or emotional problems on genetic predispositions have led to a rash of publicity over each new genetic "breakthrough." In April 1990, a lead article in the *Journal of the American Medical Association* asserted that alcoholism could be linked with a specific gene. The report was accompanied by press releases, a highly publicized news conference in Los Angeles, and video interviews with the study's authors, which the AMA transmitted by satellite as part of its weekly television news release. The *New York Times* had the story on its front page. Subsequent polls showed that over 60 percent of Americans now believe that alcoholism is genetically linked. However, after careful examination of the study, many scientists in the field came to the conclusion that the study was flawed. They felt strongly that there was no statistically significant proof of any "alcohol gene." Similar situations have occurred with schizophrenia and manic depression. Attention-getting front-page stories have been followed weeks later with back-page retractions.[40]

The new eugenics revolution will undoubtedly gain even further momentum due to the ongoing federally funded multibillion-dollar efforts to decipher the entire human genome. As scientists map, sequence, and more fully understand the function of human genes, we will hear continuing claims that various genes are responsible for a variety of human traits. Human intelligence may be the next front-page story. As we noted earlier, over a half-million dollars of federal research money is being granted for research into the genetic predisposition for high I.Q.

The human and ethical price of the new eugenics could be high. Polls already show that a significant number of Americans (11 percent) would abort a child on the basis of it being predisposed to obesity. No polls have been taken on the number of Americans who would abort a child based on its predisposition to low I.Q., alcoholism, or criminal behavior, but no doubt the percentages for each would be substantial. As we have seen, a new multibillion-dollar gene-screening industry is now developing that will tell parents the genetic predispositions of their embryos and fetuses, and another body shop industry is developing in the manipulation and engineering of human genetic traits. The age of "commercial" eugenics is here.

———

The modern incarnation of the idea that life-forms, including the human body, are machines has had extraordinary consequences. We have seen it significantly expand the human body shop, transform human work, recast our image of the human body, and lead to attempts to create more efficient future generations through a new and frightening eugenics.

As we await the final result of mechanism's war on the body, it is important to recognize that mechanism was not alone in creating the current human body shop. The engineering and commercialization of the human body is also the near inevitable result of our belief in the laws of the free market, the workings of the "invisible hand."

20

*It is only our Western Societies that quite recently turned
man into an economic animal.*
  Marcel Mauss[1]

Avarice and usury and precaution must be our gods. . . .
  John Maynard Keynes[2]

The Gospel of Greed

PAST CIVILIZATIONS ADDRESSED questions of great
public concern by attempting to ascertain what God's or the gods'
will would require. Whether making war, planting crops, or deciding
to move or stay, people in earlier times prayed and meditated for
divine guidance. Priests, oracles, and shamans—men and women
thought capable of receiving divinely inspired answers or revelations—were much relied on. Even the enlightened Greeks hastened
to the Oracle at Delphi for advice in state policy.

Today when our rulers encounter public policy dilemmas, be
they questions on employment policies, environmental protection,
health care, or even the selling of human body materials, they turn to
our modern-day oracles, the economists. When these secular shamans construct their own prophecies on vital questions of the day,
they do not rely on any religious revelation or creed, but rather on
reams of arcane statistical data and the interpretation of "economic
laws." Economist and author Robert Nelson comments, "If the priests
of old usually asked whether an action was consistent with God's

design for the world, in the message of contemporary economics the laws of economic efficiency and of economic growth have replaced the divine plan."[3]

The Golden Rule of the new economic seers is provided by the ideology of the marketplace: Do unto others, in your own self-interest. The economists' inquiries of any policy suggestion are relatively straightforward. Would a proposed activity or policy inhibit or interfere with market forces? Would it impede an individual's ability to contract for goods and services? If the answer is yes, the policy is generally undesirable; if the answer is no, it is acceptable. A second test reflects a national or global frame of reference: Does the proposed action or policy serve to advance the overall economic efficiency and the long-run productivity of the national (or world) economies? If yes, it is declared good, if no, bad.

When subjecting the important policy questions of our time to a market analysis, some economists attempt to maintain that their actions are "scientific" and therefore nonideological. They see themselves as neutral judges applying the iron-clad laws of economics. However, most economists today recognize and even relish their role as the ultimate arbiters of values and social behavior. They openly maintain that market prerogatives such as self-interest, autonomy, economic freedom, efficiency, competition, and growth will produce more social good than the outdated ethical and religious norms of past societies and cultures.

For many in the economics field, the market ideology has become "the faith for those who have no faith." One writer has noted that since the Enlightenment, "Economics was no longer a lesser branch of theology. Instead theology at this point assumed an economic content."[4] Some economists even speak in near religious terms about their discipline. Nobel Prize–winning economist Herbert Simon has observed that the "classical theory" of economics, based on the free interplay of individual self-interest, yields a condition of "omniscient" rationality, a quality formerly attributed to God. He goes on to describe the economic system as "strikingly simple and beautiful."[5] Other scientists dream of "a golden age of bliss in which nobody would have to suffer or grow hungry any longer." This golden age, which earlier philosophies and religions set in the distant past or in a future heaven, is now moved to sometime "in the near future" right here on earth.[6] The world will become "free" by the unfettered working of the marketplace.

The religion of the market has also trickled down into the general population. For most of us, economic commands have taken the place of divine intent even in our personal lives. The ability to eliminate evil in our lives is no longer thought of as a personal or community plea to the divine, but rather as a function of accumulating capital and obtaining more consumer goods. As summarized in a popular Wall Street motto, the current meaning of life is: "Whoever dies with the most toys wins." Most of us have come to believe that if market forces could eliminate want, then our personal ills and those of humanity could be healed and the world would live in basic harmony. Our contemporary usage reflects this secularization of our personal and social goals. In modern times, "full faith and credit," "trust," "value," "save," "goods," and "bond" are primarily economic terms, not personal or sacred.

The ideology of the market has been and continues to be a central operating principle of the human body shop. As the doctrine of mechanism and advances in science and technology transformed the body into machinelike spare parts, so the market "mechanism" provided the ethical basis that allowed for the sale of those parts to take place. We have seen how free-market adherents, body-shop entrepreneurs, and policymakers have successfully established the marketing of humanity—the commercialization of blood, the sale of organs in countries around the world, the open sale of elements of human reproduction, the contracting of childbearing, the marketing of fetuses and children, the aggressive peddling of human biochemicals, and the patenting of human genes, cells, and other body subparts. We have seen how courts have allowed the sale of body parts, gestation by contract, and even redefined the body as a parts "factory."

As the market faith invades the body, key questions emerge: How did the market gain dominance over other moral and ethical systems in our society? How did it become the mantra of our policymakers? How is it that the market is seen as the "moral" and legal basis for the selling of body parts, the contracting for children, and the patenting of human genes and cells? If we are to come to terms with the human body shop, and create policy strategies to limit its reach, we will have to answer these questions.

The Invisible Hand

Markets have been part of human behavior for millennia. Humans have traded with one another since the last Ice Age. From the villages of Africa to the towns of western Europe, people have always bartered and exchanged goods. There is evidence that the Cro-Magnon hunters of the central valleys of France obtained shells in trade from the Mediterranean. In northeastern Germany archaeologists discovered an oak box with a leather shoulder strap that contained Bronze Age manufactured tools, including a dagger, sickle head, and needle. Experts conjectured that the box was the sample kit of an early traveling salesman attempting to make exchanges and garner orders for the specialized products of his community.[7] Of course, the "economic" life of our ancestors included more than trade. It also involved a fair share of hunting, fishing, exploration, piracy, and invasion.

Though trading was an important adjunct to early societies, the market was never the means by which these societies solved their basic economic problems. The allocation of a community's resources among different uses, and the distribution of goods within a community, happened outside of the marketing process. The societies of antiquity never saw the market as an autonomous entity that could or should exist apart from the social strictures of any given culture. Markets and economic life in general always obeyed the religious rules and taboos of society. As noted by economist Karl Polanyi, "Never before our own time were markets more than accessories to economic life."[8]

An evocative example from a market in central Africa illustrates the point:

> Every injury occurring on the marketplace and involving the shedding of blood necessitated immediate expiation. From that moment no woman was allowed to leave the marketplace and no goods might be touched; they [the goods] had to be cleansed before they could be carried and used for food. At the very least a goat had to be sacrificed at once.[9]

Even in Western society, market forces played only a limited role compared to the cultural and religious beliefs, traditions, and social patterns that governed the lives of our ancestors. In the mid-eighteenth century, however, all this changed. Within one generation the concept of market was transformed from the literal place of trade and barter to an ideology upon which an entire social system was created, and on which central assumptions about human nature were based.

The market ideology began as an outgrowth of the scientific and technological upheavals of the seventeenth and eighteenth centuries. Imbued with optimism about the new age of productivity and invention, the Enlightenment thinkers were committed to discovering rules of human behavior that were scientific and not based on religious dogma or ancient taboo. They created the market philosophy in a self-conscious attempt to codify principles for human conduct as efficient and predictable as the mathematical laws discovered for the physical universe. Eighteenth-century philosopher Francis Hutcheson was an early pioneer in attempting to find a "physics of society" comparable to the physics of nature discovered by Isaac Newton and others. While Hutcheson did not deny the power of the traditional Judeo-Christian concept of benevolence as a principal mover of people, he posed another social force of even greater power, self-interest. Self-interest, according to Hutcheson, was to social life what gravity was to the physical universe: "Self love . . . is as necessary to the regular State of the Whole as gravitation."[10]

It was left to Hutcheson's student, Adam Smith, to transform self-interest into a revolutionary new social doctrine. For Smith, self-interest was a principle of human behavior broad enough to provide the basis for a new economic order. The eighteenth-century Scottish economist and philosopher, who is still seen today as the patron saint of modern economics, created what came to be called the "free market" doctrine. Smith preached that self-interest and market principles were an "invisible hand" that led almost magically to the good of all.

Smith's economic principles were created in part as a response to the industrial boom that emerged in his time. In 1776, the year Smith published his famous treatise on economics, *The Wealth of Nations*, England was in the midst of a technological revolution. A variety of eighteenth-century inventions, such as the flying-shuttle, spinning frame, and spinning jenny, were transforming the textile industry. These inventions resulted in a twelvefold increase in cotton consumption in England between 1770 and 1800. The invention of the steam engine, and corresponding advances in the use of chemicals and in metallurgy, also instilled a growing optimism about massively increasing the rate at which natural resources could be exploited and products created. With all the economic growth he saw around him, Smith felt that there would be little limit to the wealth that humankind could create if only left free to do so.

As spelled out in Smith's famous work, the free market dogma holds that individuals, freely pursuing their own selfish desires, are nevertheless bound to serve the common good, whether they consciously intend to or not:

> Every individual is continually exerting himself to find out the most advantageous methods of employing his capital and labour. It is true that it is his own advantage, and not that of the society, which he has in view; but . . . it is plain that each, in steadily pursuing his own aggrandizement, is following the precise line of conduct which is most for the public advantage.[11]

For Smith, the government that governs least, governs best:

> It is thus that every system which endeavors, either by extraordinary encouragements to draw towards a particular species of industry a greater share of capital . . . or by extraordinary restraints to force from a particular species of industry some share of capital . . . is in reality subversive of the great purpose which it means to promote. It retards, instead of accelerating, the progress of the society towards real wealth and greatness. . . . All systems of restraint being thus completely taken away, the obvious and simple system of natural liberty establishes itself of its own accord. Every man . . . is left perfectly free to pursue his own interest his own way, and to bring both his industry and capital into competition with those of any other man or order of men. The sovereign is completely discharged from . . . the duty of superintending industry.[12]

Smith's laissez-faire philosophy, and that of his followers, holds that if the government keeps its hands out of economic affairs, the natural selfish order of each person will speedily flower into economic well-being for all. As each pursues his or her self-interest, the "laws" of supply and demand will be applied to all commodities and will govern price and production. Through competition and self-interest society will be led to a cornucopia of economic wealth. Community obligations, or duties to one another, imposed by the church or governments, had no purpose in Smith's scheme. Such imposed duties or obligations were short-sighted, not grasping the "omniscient" benevolence of the invisible hand of self-interest.

Smith's free market doctrine was also to give rise to one of the most important tenets of the market faith. Namely, that humans had a natural "right" that allowed each individual unrestricted access to markets and natural resources. No government could interfere with this right by restricting unfettered competition, movement of workers, shifts of capital, or sale of land. And what was salvation, according to this new market dogma? According to Smith, "Consumption is the sole end of all production." Believing that men and women are basically egoists in pursuit of economic self-interest, Smith's doctrine subordinates all of human desires and needs to the quest for material abundance to satisfy physical needs. In his system there are no ethical choices to be made, only utilitarian judgments exercised by each individual pursuing his or her own material self-interest. The new divinely ordained purpose of society is to consume, and to produce ever more goods for consumption. The human being, after Smith, has been dubbed "Homo Consumptor."[13]

Smith's teachings would sustain the growth of the Industrial Revolution and provide the moral basis for the development of technology and the factory production system. Smith's market theories, one historian notes, "evolved into a veritable faith in man's secular salvation through a self regulating market."[14] Economist Robert H. Nelson sums up the revolution in theology and ethics accomplished by Smith and his peers:

> If the current sinful state of human affairs involves the coercive exercise by some human beings of power over others, economists offer mankind a way to a future world in which all relationships will be based on voluntary consent and a perfect harmony will have been achieved. In short, to encourage self-interest is not to

> encourage divisions and coercive measures within humanity; it is instead to establish a necessary condition for reaching an existence of perfect equality and rationality—essential elements in reaching the future heaven on earth promised by modern economic progress.[15]

Smith's message is still a central secular theology today, as we are told that a new world order, founded on international competition and free trade, will lead to endless production of wealth, the spread of democracy, and the triumph of human liberty. "Capitalism," Milton Friedman assures us, "is . . . necessary for political freedom."[16]

Few today have read Adam Smith, yet the gospel of self-interest has become second nature to most. Greater productivity, individual self-interest, and upward mobility (financial, not spiritual), the glorification of competition, greater consumption of goods and services, and suspicion of government interference with business affairs remain unquestioned idols in our society. They are as firmly entrenched in our times as any church dogma was in the Middle Ages. History has shown Edmund Burke remarkably prescient when he commented, shortly after the *Wealth of Nations* was published, that "In its ultimate result, this was probably the most important book that had ever been written."[17]

Market Fictions

For many, society's conversion to the free market faith has been successful. More than any prior ideology or theology, it seems to have brought an earthly heaven within reach. Much of our current wealth and technological achievement are credited to the profit incentives provided by the market system. And now, as the market ideology begins to treat the human body on a supply-and-demand basis, market adherents argue that the consequences will once again be beneficial. They allege that open sale of body parts will provide for more organs at lower prices, fairer allocation of semen, eggs, children to those who wish them, and greater incentive to develop useful products from genetically engineered human genes, cells, organs, and embryos. Those opposing unrestricted buying and selling at the human body shop are viewed as reactionaries who still do not see the benevolence of the invisible hand.

But market advocates show a selective amnesia about the impacts of the invisible hand. They do not recognize that the birth of the laissez-faire market system administered a shock to our view of nature and ourselves from which we have not yet recovered. Nor do they appear to understand that there was a problem with the market system from the very start.

The vision of Smith and his followers, of a society of freely contracting individuals, autonomous and self-seeking, was centrally flawed. For a contradiction exists at the very heart of the conception of the universality of market laws, and the establishment of a society on the "physics" of self-interest. If the market system is to work, if supply and demand are to be quasi-religious laws free of government regulation, then everything has to be for sale. Everything has to be a commodity in the market. Every element of industry, each aspect of production—machines, labor, land, other natural resources, money, goods—is regarded as having been produced for sale in the market and therefore subject to the supply-demand price "mechanism." These numberless markets are then, under the theory, interconnected and form one overarching market system.

However, not everything is a commodity. In the economic context, the word commodity has a limited and precise meaning. Commodities are defined as manufactured goods produced for sale. Commodities, be they clothes, cars, or computers, are produced by people, sold, and eventually consumed. That is their origin and purpose. Clearly, central aspects of any society do not fit the definition and purpose of commodities.

For example, human labor is not a commodity. It is not manufactured for sale and consumption. It is both artificial and misleading to attempt to neatly package and commodify labor, thereby separating wage work from the rest of the life of each of us. Labor is not a product—a watch or motor—but rather a personal, intimate, and intrinsic part of ourselves. Human work cannot be separated from the whole person. Whoever has purchased labor has purchased not just the work of an individual, but also a significant and indivisible portion of the life, thoughts, and creativity of the worker—their full presence. For the hours of the work day or work night, the employer controls not a working "machine," but the environment and well-being of a person. Moreover, the buying and selling of labor determines far more than just how and by whom labor will be done. It affects central aspects of a worker's life, including where the worker and the worker's family will live and how they will live. When searching for buyers of their labor, workers know that not just their work but their societal worth and future and the well-being of their families are on the line. In sum, it is a market fiction that there is a separation between the human and human work. We can no more sell our work than we can sell ourselves.

Land is also not a commodity. Land is just another name for a part of nature. Land is, of course, not produced by people for sale and consumption. It is a given, a gift. Land has intrinsic worth and meaning that can never be measured by the reduced concept of marketplace value, determined by supply and demand.

The noncommodity nature of key aspects of any society or industry, including labor and land, presented a central challenge to the advocates of the market system. If market ideology was to be the central law of a society, higher than religious or cultural traditions, it had to extend to all important aspects of social life. Work and nature could not be left out of the market equation. They could not be excluded from the system and controlled by tradition, cultural values, or other means of social organization, for then the whole system would become nonviable. Vital noncommodities had to be subsumed under the definition of commodity, treated like any other commodity, and subjected to the supply-and-demand laws of commodities, no matter how irrational this appeared.

By a breathtaking philosophical maneuver, market proponents have for over two centuries simply ignored the distinction between commodities and noncommodities. They have created the fiction that elements of human society and nature, such as labor and land, were commodities to be sold and consumed. The bold sleight-of-hand by which key "fictitious commodities" were created gave the market system control of virtually all aspects of social behavior and natural resources. This allowed the market doctrine to obtain political and philosophical hegemony over Western society. As noted by Karl Polanyi:

Liberal economy, this primary reaction of man to the machine, was a violent break with the conditions that preceded it. A chain-reaction was started—what before was merely isolated markets was transmuted into a self-regulating *system* of markets. . . . The crucial step was this: labor and land were made into commodities, that is, they were treated *as if* produced for sale. Of course, they were not actually commodities, since they were either not produced at all (as land) or, if so , not for sale (as labor). Yet no more effective fiction was ever devised. . . . Accordingly, there was a market price for labor, called wages, and a market price for the use of land, called rent . . . The true scope of such a step can be gauged if we remember that labor is only another name for man, and land for nature. The commodity fiction handed over the fate of man and nature to the play of an automaton [the free market] running in its own grooves and governed by its own laws.[18]

The social history of the last two centuries has in many respects been the result of the contradiction and tensions inherent in the market system's creation of fictitious commodities. Over time, treating certain noncommodities as commodities became a double-edged sword. On one hand it led directly to massive increases in wealth, technological development, and consumption, as modern market economies spread over the face of the globe. But it also led to the downfall of the pure laissez-faire market system, as the "invisible hand" showed itself capable of causing very visible havoc. The selling of human labor, nature, and other fictitious commodities (such as money) precipitated innumerable social abuses, dislocations, and disasters—child labor, inhuman working conditions, impairment of health, destruction of families and communities, uncontrolled exploitation of nature, extinction of species, air and water pollution, wildly fluctuating economic conditions, depression and inflation, spiraling governmental and personal borrowing. At various times these profound problems have threatened the very cohesion of society, even imperiling the sustainability of the biosphere.

The societal and environmental crises brought on by the market system's commodification of noncommodities has forced governments throughout the world to abandon the laissez-faire free market and to step in to limit and occasionally eliminate the market system in various areas. These efforts to shore up the fissures in the free market structure have, more often than not, resulted in vast and unwieldy bureaucracies which attempt to regulate and control the ways in which fictitious commodities (i.e., labor, land, and money) are sold. Virtually every industrial capitalist nation has a vast infrastructure of agencies set up to regulate wages, ensure worker safety, provide for the temporarily unemployed, grant social security for those unable or too old to sell their labor, guarantee health coverage, protect various land through zoning or the creation of parks and wilderness areas, limit industrial pollution, regulate interest rates, provide insurance to bank depositors, etc. In some cases socialist and communist countries have taken full control of a society's means of production and eliminated the free market system altogether. However, these "state capitalist" societies have continued to treat labor and land as exploitable commodities as part of their attempt to compete (most often unsuccessfully) in the global marketplace.

Ironically, free market advocates now aggressively advocate against regulatory bureaucracies and the managerial elites which have risen up in many countries, especially in socialist systems of economic control. These critics of government bureaucracy and "managerialism" correctly point out the oppression and inefficiencies built into the modern technocratic state. Yet they refuse to examine the origins of these mammoth regulatory instruments. They continue to extol the virtues of the free market while not recognizing that the government control and regulation which they oppose was made necessary by the dislocation caused by the free market itself, as it unsuccessfully attempted to treat noncommodities as commodities.

The turbulent history of the free market and fictitious commodities is one of key relevance to the controversy surrounding the human body shop. For as described, we are now in the early stages of adding the human body, its parts and processes, to the list of commodities that are subject to the laws of supply, demand, and price. The body is not a commodity. It is not a manufactured product intended for consumption. However, just as the new techniques in industrial technology led to the commodification of noncommodities such as human work and nature, the new techniques in biotechnology, including transplantation, reproductive technology, and genetic engineering, are now leading to the commodification of the body. In order to better understand what is in store for the human body as the newest ficti- tious commodity in the marketplace, it is instructive to undertake a brief review of the history of prior commodity fictions, especially that which is almost as intimate to us as our own bodies: human labor.

21

A man's labor also is a commodity, exchangeable for benefits as well as any other thing.
Thomas Hobbes[1]

Satanic Mills

AT THE OUTSET of the industrial revolution, the sale of labor was the most important and visible noncommodity treated as a commodity. The commodification of labor was made possible by the growth of the factory system. Before mechanization skilled workers sold many of their products, but not their labor. The late eighteenth-century textile industry was the first to witness the transformation (some would say degradation) of labor, through the displacement of human weavers by machines. The new textile factories replaced individual and family workshops with factories and wage work. Mass-produced products began to replace those fashioned by individuals. The increased use of power and machinery in the textile industry signaled the beginning of a revolution in the ways human work was seen and organized. As historian Lewis Mumford notes:

> It [mechanization] marked the end of the guild system and the beginning of the wage worker. It marked the end of internal workshop discipline, administered by masters and journeymen through a system of apprenticeships, traditional teaching, and

the corporate inspection of the product; while it indicated the beginning of an external discipline imposed by . . . the manufacturer in the interest of private profit. . . . All this was a large step downwards. In the textile industries the descent was rapid and violent during the eighteenth century. In sum: as industry became more advanced from a mechanical point of view it at first became more backward from a human standpoint.[2]

Early advances in industrializing the manufacturing of textiles were a threat to many local communities that survived on the premechanical production of cotton. In 1780, the story goes, Ned Ludd, a boy from Leicester, England, expressed his anger toward mechanization by demolishing some stocking frames. Thenceforth those who took up machine sabotage were known as "Luddites." Soon battles between local artisans and the new "machine shops" erupted throughout Europe. In Scotland hand loom and power loom weavers engaged in pitched battles. In France J. M. Jacquard's ingenious brocade silk looms, with their punched card system for regulating designs, brought on riots from which the inventor barely escaped with his life. Barthelmy Thimmoniers's sewing machines caused violent upheaval in Paris, and a mob destroyed his shop.

Although the new division and specialization of labor was disastrous for many, it provided Adam Smith with the basis for the concept that human work could be commodified like any other product. Smith's whole historic market theory was posited on the sale of this labor. Under Smith's theory the division of labor, and every person's ability to sell his or her labor for a wage, as a commodity, allowed for a secular salvation of society. The first sentence of Book One of *The Wealth of Nations* reads, "The greatest improvement in the productive powers of labor, the greater part of skill, dexterity and judgment with which it is anywhere directed, or applied, seem to have been the effects of the division of labor."[3]

Division of labor meant more goods could be produced in less time and at a far cheaper cost per unit. Sir William Petty, one of the new breed of economists spawned during the period, praised the application of division of labor in the construction of watches: "In the making of a watch, if one man shall make the wheels, another the Spring, another shall engrave the Dial-Plate, and another shall make the Cases, then the watch will be better and cheaper, than if the whole work be put upon any one Man."[4]

Economist and author Milton Myers notes that for many the division of labor was equivalent to "Prometheus bringing the gift of fire down from the Gods to serve the needs of man."[5] The new Promethean technologies of the eighteenth and nineteenth centuries allowed the market advocates a vision of the future in which entrepreneurs would create more and more enterprises that demanded factorylike division of labor, and in which millions of individuals would sell their labor on a supply-and-demand basis. Self-interest for manufacturer and laborer alike could now be seen as leading to a growing specialization of tasks, which in turn was visibly increasing production, output, and societal wealth.

At the time Smith was propounding his market theory, a revolution in the demographics of England had created the massive pool of laborers required to make the commodification of labor a reality. The village commons had been the primary social unit in England for several centuries. Peasants, called "freemen," pooled their individual holdings into open fields that were jointly cultivated. Common pastures were used to grow vital crops and graze animals. Village commons peppered the English countryside. They were self-governed by councils and usually demonstrated primitive forms of democratic decision making. In the early 1500s, a cynical expropriation of common land known as the Enclosure movement had begun, an expropriation that continued right through the first half of the nineteenth century. Enclosing means "surrounding a piece of land with hedges, ditches, or other barriers to the free passage of men and animals." The gentry threw up fences around fields that had fed villagers and their livestock for hundreds of years. They prevented freemen and their families from access to their livelihoods. Large landowners destroyed the centuries-old commons system to make more land available for sheep and wool production, which were gaining high export prices.

Fenced off from their land, the peasants reacted. Protests and rebellion ensued. Sir Thomas More was one of many who attacked the enclosers and their greedy expropriation based on wool prices: "Your sheep, that were wont to be so meek and tame and so small eaters, now . . . become so great devourers and so wild, that they eat up and swallow down the very men themselves. They consume, destroy, and devour whole fields, houses and cities."[6] With the phrase "Sheep devour people" being uttered on all sides, millions of peasants were dislodged from their ancestral homes and forced to migrate to the new, industrializing cities, where they became the first people to attempt to "sell" their labor in the market system. There, living under execrable conditions, they became the workers, the proletariat, available for the commodification of labor praised by Smith and demanded by the factory system. Now the factory rather than the farm became the center of social and economic life. After 1850, the factory was not only the key economic institution of England, but also the primary force that shaped its politics, its social problems, and the character of its daily life.

It is difficult for us today to realize the pace or the quality of change that this rise of factory work brought about. Until the mid-eighteenth century, many cities that were to become major industrial areas were mostly farmland or undeveloped forest. Manchester in 1727 was described by Daniel Defoe as "a mere village." Forty years later it was filled by a hundred integrated mills and a whole cluster of machine plants, forges, and leather and chemical works. A modern industrial city had been created. By the 1780s, the shape of the new environment was visible. A French mineralogist visiting England in 1784 wrote:

> [The] creaking, the piercing noise of the pulleys, the continuous sound of hammering, the ceaseless energy of the men keeping all this machinery in motion, presented a sight as interesting as it was new. . . . The night is so filled with fire and light that when from a distance we see, here a glowing mass of coal, there darting flames leaping from the blast furnaces, when we hear the heavy hammers striking the echoing anvils and the shrill whistling of the air pumps, we do not know whether we are looking at a volcano in eruption or have been miraculously transported to Vulcan's cave.[7]

The factory provided not merely a new landscape but a new and uncongenial social habitat. We have become so used to urban industrial life that we forget how wrenching is the transition from farm to city. For peasants desperately seeking to sell their labor in an industrial setting, this transfer requires a drastic adjustment. No longer do they work at their own pace, but at the pace of a machine. No longer are slack seasons determined by the weather, but by the state of the market. No longer is the land, however miserable its crop, a perennial source of sustenance close at hand, but only the packed and sterile earth of the industrial site.

Not surprisingly, the local textile operators took full advantage of the newly dislocated and impoverished population. Thousands of destitute children sold their labor as a commodity and were employed in the new factories. Adults competed with one another for bare subsistence in the new industries. The numbing work and terrible conditions became public scandals. A Committee of Parliament, appointed in 1832 to look into conditions, gives this testimony from a factory overseer.

Q. At what time in the morning, in the brisk time, did these girls go to the mills?

A. In the brisk time, for about six weeks, they have gone at three o'clock in the morning and ended at ten or nearly half past at night.

Q. What intervals were allowed for rest and refreshment during those nineteen hours of labour?

A. Breakfast at a quarter of an hour, and dinner half an hour, and drinking a quarter of an hour.

Q. Was any of that time taken up in cleaning the machinery?

A. They generally had to do what they call dry down; sometimes this took the whole time at breakfast or drinking.

Q. Had you not great difficulty in awakening your children to the excessive labour?

A. Yes, in the early time we had to take them up asleep and shake them.

Q. Had any of them any accident in consequence of this labour?

A. Yes, my eldest daughter . . . the cog caught her forefinger nail and screwed it off below the knuckle.

Q. Has she lost that finger?

A. It is cut off at the second joint.

Q. Were her wages paid during that time?

A. As soon as the accident happened the wages were totally stopped.[8]

It was a grim age. The long hours of work, the general dirt and clangor of the factories, and the lack of even the most elementary safety precautions all combined to give early industrial capitalism a reputation from which, in the minds of many people of the world, it has never recovered. Worse yet were the slums to which the majority of workers returned after their travail. Life expectancy at birth in the newly developed city of Manchester was seventeen years—a figure that reflected a child mortality rate of over 50 percent.

"Labor" became the technical market term used for human beings, in so far as they sold their work as employees. Using the general term "labor" for entire classes of humans indicated how completely Smith's ideology had triumphed. Humans were now defined by their place in the economic system. They in turn no longer worked in the interests of the community, or for family or religious duty, but rather for bare subsistence wages. Laborers' lives and the lives of their families had become an accessory of the economic system. The structure of the family, its mobility, and the relationships therein were determined by how and where each member could sell his or her labor.

The extraordinary exploitation inherent in the free market commodification and consumption of human labor led to open revolt. Workers rebelled, and their strikes and disruptions threatened the stability of society itself. The maintenance of social cohesion required regulation and restriction of the labor market. No government could withstand the excesses of the "satanic mills" treatment of workers. The necessary restriction on the labor market inevitably led to the death of the pure free market system in England. In 1802, pauper apprentices were legally limited to a twelve-hour day and barred from night work. In 1819, the employment of children under nine was prohibited in cotton mills. In 1833, a forty-eight- to sixty-nine-hour week was decreed for workers under eighteen (about 75 percent of all cotton-mill workers), and a system of government inspection of factories was inaugurated. In 1842, children under ten were barred

279

from the coal mines. In 1847, a ten-hour daily limit (later raised to ten and a half hours) was set for children and women.

A century after Smith's thesis was published, labor unions were being formed in many industrializing countries to protect workers from the horrors of early capitalist exploitation. The first decades of the twentieth century were marked by massive conflict between employers and workers over wages and a variety of workplace conditions. Subsequently, government agencies were set up to protect workers from unacceptable exploitation, and a variety of laws were passed, including those mandating a minimum wage, nonhazardous workplaces, and some economic protection for the unemployed.

Despite reform, however, the commodification of labor is still a social crisis for many. A recent headline in the *New York Times* noted that in the United States, "More Children Are Employed, Often Perilously." The article described how millions of children are still at work, many at hazardous jobs. Other reports indicate that child-labor abuses remain pervasive in U.S. workplaces.[9] Ironically, while children are being employed in the American work force, millions of adults are unemployed. They cannot sell themselves in the labor market and are condemned to attempt a livelihood at the margins of society. Even those who successfully sell their labor are having to work increasing hours as compared with the last generation. Americans, by choice or by force, now work almost 160 hours more a year—that's an extra month of work—than they did twenty years ago. Americans work 320 hours a year more than workers in such countries as Germany and France and are given only one-third the days off that many Europeans get. Moreover, as noted earlier, millions of Americans commit workplace *hara-kiri*, working themselves into a variety of work-related diseases and injuries. Currently, our modern industrial and postindustrial workplaces are beginning to take a toll similar to the so-called satanic mills of earlier times.[10]

Reviewing the history of the sale of labor as a fictitious commodity, it becomes evident that the marketing of labor was a precursor to today's human body shop. The selling of a lifetime in the form of hourly wages is only one step removed from the contract sale of lives, or the renting of wombs through surrogate parenting contracts. The same principle that allowed for the sale of a person's most precious possession—time might also allow for the sale of one's precious bodily elements—blood, organs, semen, or ova. Moreover, if one can patent machines and useful mechanical inventions why not declare patentable those other useful living "inventions," including humans, which are also sold and contracted for in the industrial system? If we can alienate humans from their bodily labor why not also from their bodies?

The Earth's Body

When we contemplate the future of the human body shop, we cannot ignore the crises created by the commodification of land. For land, like the body, is a gift (or at least a given), not a commodity or manufacture. Yet land was also successfully invaded by the market, and the commercialization of land has had an enormous and frightening impact. The most pressing internal contradiction in the market system today is the global environmental crisis. It is becoming increasingly clear that current problems in keeping our biosphere sustainable are not just the by-products of mismanagement or poor judgment, but rather the direct outcome of the market mode of production brought to us by Smith and his predecessors, including John Locke.

Locke, the most influential English thinker of the seventeenth century, died nineteen years before Smith was born but had a profound influence on the Scottish philosopher. Locke's idea of secular progress closely resembles Smith's. Locke looked on nature as a vast, unproductive wasteland. His view of "value" meant that the natural world was only of worth when human labor and technology transformed it into useful commodities. The more quickly that land could be transformed into a store of material goods, the more secure a society would become and the more progress civilization would achieve. As stated by Locke, "Land that is left wholly to nature . . . is called as indeed it is, waste . . . on the other hand, he who appropriates land to himself by his labor, does not lessen but increase the common stock of mankind."[11]

Locke's view was revolutionary. For centuries English land had not been marketed. It was handed down through families or kept by the church, but strict laws forbade its being sold. However, the market doctrine provided the rationale for the unchecked exploitation of nature and the view of land as a commodity like any other. It denied land any inherent noncommercial value. Locke's transformation of nature from a good into "goods" radically reoriented our relationship to nature. Suddenly, the natural world—including humans —was not a community of subjects, but rather a collection of exploitable objects. When we consider a whole new category of global environmental problems—ozone depletion, acid rain, the "greenhouse effect," undisposable radioactive waste, species extinction, deforestation—it is clear that we are just beginning to assess the bill that the legacy of the Enlightenment thinkers and the market system have imposed on future generations.

As with labor, government has tried to restrict the free market commodification of the earth in order to avoid ecological catastrophe. For decades zoning laws, the creation of parks and wilderness areas, and numerous other legislative controls on the market have attempted to put some U.S. land off-limits to the market. Additionally, in the early 1970s, a group of statutes were enacted by the U.S. Congress to protect land from toxic dumps, and to secure a measure of protection for our water and air. Most European countries have passed similar environmental laws. Finally, as the threat of global environmental degradation increased, international summit meetings have been held in order to begin the process of changing the current industrial destruction of the biosphere. The attempts to reverse and ameliorate the destruction caused by the free market exploitation of the earth have met with some results, but the fate of the earth is far from assured. More needs to be done, and done quickly.

Just as the market over centuries enclosed essential fictitious commodities—the labor of humans and the earth itself—it is now attempting to enclose our bodies, our blood, organs, reproduction, even the genes, which literally are our common heritage. As we have seen throughout this book, the commodification of the body, like that of labor and land, is causing significant social dislocation and suffering. Tens of thousands of poor people around the world are selling their organs. Fetal tissue is being bartered and collected. Economically disenfranchised women and men are selling their reproductive essences. Childbearing and children are being contracted for. Women's bodies and psyches have become the laboratories of the flourishing business of reprotech. Animals are being genetically mutated, often cruelly, in order to make them better products or factories for the production of valuable human genes and biochemicals. And corporations are posed to gain patent ownership of all human genes and valuable cells.

Despite our growing regulatory bureaucracies, we have yet to find successful ways to restrain our market-driven technologies from consuming our lives through commodified labor and our earth through expropriating land and resources. As we assess the price of market ideology in our work lives and in the environment, and attempt to arrive at alternative sustainable modes of treating work and nature, we must also devise ways of dealing with the new body technologies so that our very beings do not continue to be new market commodities. We must find principles and policies that reassert the moral regard for the body effaced by the market so many generations ago. We must fashion alternatives to the market view of ourselves. If not, we can look at the satanic mills of the early industrial age and the current exploitation and destruction of the earth as models of what we will do to our bodies.

22

Our present definition of technology may be too narrow.
The paradigm shift that instigated the bio-revolution, in
effect, recognized life as technology. Viewed from this
angle, the distinction between man-made machines and
living one blurs. In fact they become one and the same.
Sharon and Kathleen McAuliffe, Life for Sale[1]

What I see coming is a gigantic slaughterhouse, a
molecular Auschwitz, in which valuable enzymes,
hormones, and so on will be extracted instead of
gold teeth.
Dr. Erwin Chargoff

At the Crossroad

TODAY WE ARE at a dramatic crossroad in our
treatment of and thinking about the human body. We can choose
to continue in our blind trust of technology and the invisible hand
of self-interest economics. We can continue to adhere, in near religious fashion, to the centuries-old dogmas of mechanism and the
market. We can continue to view our bodies as machines and commodities. We can continue to remake our bodies with surgery and
genetic engineering. We can continue to manipulate the reproductive
process by eliminating the birth of children with undesirable traits.
We can continue to alter with drugs and genetic therapies those with
"abnormal" traits. We can continue to clone life-forms and human
body parts. We can continue to permit the international sale of organs, the commercialization of fetal parts, the sale of sperm and eggs,
and the patenting of animals and human genes and cells. This is
the course that, with a few exceptions, our governments, regulatory
agencies, scientific review boards, and courts have decided on.

We can also decide to do nothing, hoping that the experts will
come to a rational consensus about the body shop and implement

sound limits to technology and the marketing of life. However, history shows that our scientists and technologists have virtually never limited a technology to its beneficial purposes. The nuclear revolution gave us X-rays, but also brought us to the brink of global destruction. The petrochemical revolution brought us a massive increase in agricultural production, industrial output, and transportation mobility, but also led to species extinction, top-soil erosion, acid rain, ozone depletion, global warming, and myriad other environmental threats. The lesson of the past is clear: Unless human choices control technology, technology will control human choices. The body shop is no exception. The new medical and genetic engineering technologies of the last few decades have and will save lives, and perhaps will provide new cures for humanity; but without appropriate limits these technologies and the market ideology behind them will also lead to the devaluation and commercial exploitation of the body.

The market, in concert with advances in technology, has also brought us extraordinary productivity. But the market, like technology, has not been limited to beneficial uses. Uncontrolled, it has led to unacceptable exploitation of people, other life-forms, and the earth itself. Nor has the market curbed our abuse of industrial, chemical, or biological technologies. The market did not limit or restrict pollution, nuclear proliferation, abuses in labor, or the sale of human parts. Left unregulated, the market does not control technological abuse; it encourages it.

If we do not shut down the body shop, our future course is clear. As we have seen throughout this book, the technologies for the full-scale commercialization of human parts are in place, and biotechnologies are becoming more sophisticated every day. Additionally, enormous profits on body elements and processes are awaiting the scientists and biotechnologists who are pushing the new body technologies. Unchecked, these advances in biotechnology will significantly accelerate current body-shop trends. Increasing efficiency in organ transplants will cause growing shortages in available organs. This greater demand will create extraordinary social pressure to increase organ supply by expanding the definitions of death to include people with higher brain loss, and also to create profit incentives for organ donation through an open or controlled market in organs. Advances in utilization of fetal parts for curing disease or for enhancement therapy will make the unborn ever more valuable commodities for the medical marketplace. As reprotech becomes refined and increases its percentage of successful births through artificial insemination, egg donation, and embryo transfer; sperm, eggs, embryos, and children will be subject to more intense marketing. As genetic screening of the unborn allows prospective parents to know a wide range of genetic traits of their offspring, eugenic abortions based on sex and other nondisease characteristics will increase. Greater understanding of the location and functions of genes will accelerate the marketing and patenting of valuable genes. Discoveries of genetic links to undesirable physical or social traits will lead to continuing genetic manipulation of humans, including proposals to permanently change the genetic code of certain individuals through germline genetic surgery. Genetic engineering will also allow for ever more transfer of human genes into animals, animal genes into plants, and plant genes into foreign species. Finally, breakthroughs in cloning could change the modes of reproduction for all members of the living kingdom, including the reproduction of humans.

Choosing to continue the course charted by the human body shop will have profound social and legal repercussions. The tragic history of the eugenic movement over the last century, and its current reincarnation in sex selection abortions, genetic screening for insurance and employment purposes, and the genetic engineering of humans, animals, other living creatures, remind us that traditional protections for humans and for genetic diversity break down under the ideology and technological onslaught of the body shop. The crises precipitated by the market's invasion of human labor and the earth showed that cultural and legal limits on how we treat humans and nature can be overthrown within a few generations. Now the human body shop is refashioning traditional concepts of birth, life, and death, of what is worthy of life and what is not.

The engineering and marketing of the body and its parts is not inevitable. We can choose another path. We can declare a truce with our physical selves. We can cease to treat our bodies under the forced regime of mechanistic efficiency and recognize them as intrinsically good despite and perhaps because of their diversity and fragility. We can view them as gifts from a creator or from millennia of creative evolution, gifts that cannot and should not be bartered and sold.

To do this we will have to reverse the current theology that determines human body policy. We will have to fight the reigning dogmas of mechanism and the market. This will not require that we abandon transfusions, transplants, or research on reproduction and the human genome. Nor will it mean limiting how much we learn about our biological selves. However, it will require a strong challenge to many of the basic assumptions on which our current economic and social system is based. It will demand the courage to say that just because a technology can be implemented does not mean it should be implemented. It will require the realization that in a civilized society some things just cannot be for sale. Are the people of the United States and other nations ready to take this path, to initiate what might be termed a "body revolution"?

Where We Stand

As we confront the epochal crossroad in our treatment of the body, American and international public opinion remains split on which path to take. Numerous polls and surveys indicate that, as of now, a majority of the public is still opposed to the full-scale marketing of humanity, clinging to a sense of the body as sacred and inviolable, despite the vertigo induced by the whirlwind of advances in the engineering and selling of body parts. Nevertheless, evidence shows that a growing minority, including a majority of the younger generation, are willing to sanction the marketing of the body.

This pattern is especially true for the sale of organs. A majority of Americans still reject the sale of human organs. According to a 1991 *National Law Journal* poll, Americans, by almost a two-to-one margin, strongly oppose the idea of a market in transplantable organs. Sixty percent say it should not be legal to pay people to agree to give up organs after death, and 63 percent say it should not be legal to get paid for giving up—during life—any nonessential organs such as a second kidney or eye.[2] Of those sixty-five years and older, only 22 percent said yes to payment for postdeath organ donations, and 25 percent said yes for organs relinquished during life.[3]

However, the poll also demonstrated that younger adults are apparently more prone to extending the market ideology into the body than are older people. Close to half (44 percent) of those polled who were in the eighteen to twenty-four age group said it should be legal for people to get paid for nonessential organs given up during life or vital organs given up after death.

The same pattern held true for contract motherhood. While Americans are split down the middle on the legality of couples paying a surrogate mother to conceive a child using the sperm of the husband, only 27 percent of those sixty-five years and older polled said they approved of paid surrogacy, while 65 percent of those eighteen to thirty-four approved of the contract childbearing arrangements. Interestingly, according to the poll, African-Americans are far less supportive of surrogate parenting than are whites. Only 30 percent of all African-Americans, regardless of age, support surrogacy, as opposed to 54 percent of whites. Professor Patricia King, an ethicist expert at Georgetown University, explains the difference: "The black community is conservative, and traditional, and surrogacy is a real break [with the norm]."[4]

Young people are also more willing to see patients profiting from their human tissues. When confronted with a scenario similar to the *Moore* case (described in chapter 15), 62 percent of people eighteen to twenty-four years of age felt that a patient should get a share of profits developed from their tissue or blood. Commenting on the series of responses seeming to show a growing willingness by the young to allow the selling of body parts, ethicist Franklin M. Zweig commented, "You may be witnessing a burgeoning interest in the market-oriented approach to human parts." He added, "It will be fascinating to see if the increasing success of medical technologies furthers such a market."[5]

Other poll results show more consistent support among all ages for protection of fetal life. Those surveyed were given a scenario in which a divorcing couple was litigating over the custody and disposition of frozen embryos, a fact pattern identical to the *Davis* case (discussed in chapter 7). Sixty-four percent of the American public said that the embryos should be given to the ex-spouse who wants to become a parent; 17 percent say the embryos should be put up for adoption. Only 24 percent say the embryos should be destroyed (the position maintained by the Tennessee Supreme Court in the *Davis* case).

Moreover, a recent national magazine survey showed significant public opposition to medical use of aborted fetuses. Sixty-one percent of the thirteen hundred respondents felt that aborted fetuses should not be collected and used for research. Additionally, 79 percent felt it would be wrong for a woman to become pregnant in order to donate the tissue to save the life of a fatally ill relative.[6]

A slim majority of Americans also wish to fully retain the traditional social and legal concepts of motherhood.[7] If a woman is implanted with an embryo, should she (the birth mother) have custody rights, or should she be stripped of her parental rights and be legally forced to relinquish the child to the genetic parents? When confronted with this hypothetical situation similar to that involved in the Anna Johnson case (described in chapter 8), 43 percent of Americans felt that the birth mother of a child who had been implanted in her as an embryo should get custody, while 39 percent would give custody to the genetic parents. (Up to this point, the California courts have sided with the genetic parents, denying Anna Johnson, the birth mother, any legal rights whatsoever).[8]

In some of the cutting-edge controversies of genetic engineering, polls show that a large majority are not willing to allow full-scale use of biotechnology. According to the polls, the public does not wish to see the genetic integrity of humans or other species compromised. A 1992 survey by the U.S. Department of Agriculture on genetic engineering showed that when confronted with some of the transgenic creations discussed in chapter 13, most of the public said "no." When asked the question, "Would [a] chicken be acceptable or unacceptable if [some of its] genes came from a human?" only 10 percent of those polled found such a chicken acceptable. When asked whether potatoes with genes from an animal would be acceptable, only a quarter of those polled thought that it was. When asked if a chicken made less fatty by the insertion of genes from another animal was acceptable, almost 60 percent opposed this type of genetic engineering.[9] Overall, over 50 percent felt that using biotechnology to change animals was "morally wrong." Interestingly, most felt that the public should have a greater voice in biotechnology decisions; 79 percent believing that "citizens have too little a say in decisions about whether or not biotechnology should be used."[10]

Poll results on the appropriateness of the genetic engineering of humans are more ambiguous. In September 1992 the March of

Dimes surveyed one thousand U.S. adults on gene therapy and testing. At the outset the poll found limited basic knowledge of genetic engineering in the general public. Over two-thirds of those polled knew "relatively little" or "almost nothing" about genetic testing, Almost 90 percent knew little or nothing about gene therapy. Nevertheless, 79 percent say they are willing to undergo gene therapy to correct a serious or fatal disease. An even larger percentage (89 percent) would be willing to submit their child to gene therapy to cure serious illness. While the majority disapproved of use of gene therapy for cosmetic purposes, a surprisingly large number (43 percent) would approve of its use to enhance physical characteristics, including 42 percent who would use gene engineering to improve the intelligence level children would inherit. Those polled were generally concerned about the possible abuse of gene therapy. Nearly three-quarters believe that "strict regulations" are necessary.[11]

Questions about genetic testing elicited similar results. Almost eight out of ten people surveyed said that they would take genetic tests before having children, to discover whether their future offspring would be likely to inherit a fatal genetic disease. Seventy-two percent say they would take genetic tests to determine whether they or the children they already have were likely to have a serious or fatal genetic disease. Almost two-thirds say that they would have genetic testing during pregnancy to determine serious disease. As for confidentiality, two-thirds of those polled felt that an employer should not know the genetic predisposition of an employee. [12]

Gene patenting does not fare as well in public opinion as genetic screening or therapy. International polls indicate that an overwhelming number of the public, and even of scientists, oppose the patenting of genetic material from humans.[13] A majority of the public also opposes animal patenting.

At present, significant percentages of the public in the United States and elsewhere seem ready to close down the human body shop. Many, however, including the younger generation, are leaning toward allowing the commodification of the body. At this crucial juncture it is incumbent on advocates of the inviolability and integrity of humans and other life-forms to present a different vision of the body and society than that promulgated by the proponents of the body shop. As it stands at the crossroad, the public needs clear guideposts to steer it through the age of biology in a way that protects the human body from the incursion of mechanism and the market.

The final chapter of this book is an attempt to outline a manifesto of sorts for those seeking to protect the body. It provides ethical views which offer a stark contrast with those currently in place in the human body shop. It also lists a series of biopolicies for the future—policies based on viewing the body empathetically rather than mechanistically, and viewing the body as gift rather than marketable merchandise.

23

Any religion which is not based on a respect for life is not a true religion.
    Albert Schweitzer[1]

Unless we change direction, we are likely to end up where we are headed.
    Chinese Proverb[2]

The Body Revolution

CLOSING DOWN THE human body shop will require a counterrevolution against the current invasion of the body by technology and the marketplace. This revolution has a vanguard in the holistic health, natural childbirth, natural foods, and environmental movements, and in other efforts to develop respect for human life and the environment. It has firm support in those legislatures, in the United States and around the world, that have declared the sale of organs, whether adult or fetal, illegal. It is bolstered by the courts, and countries, that have outlawed surrogate motherhood. It is supported by nations that limit genetic screening of the unborn to questions of serious and life-threatening diseases. It is underscored by those countries (not including the United States) that refuse to allow the sale of sperm or eggs and that do not treat embryos as property. It is epitomized in those laws which prohibit the patenting of life-forms.

Every revolution is in part a revival. And the body revolution needs to revive the understandings of numerous religious traditions that have always understood the body as meaningful and sacred. The Judeo-Christian tradition is straightforward in its general rule—human beings should not be used as mere means to ends. The congressional Office of Technology Assessment did a survey of the Western religious traditions' view of the human body, in order to help the U.S. Congress better legislate on the issues of ownership and sale of the human body. Its conclusion:

> Religious traditions offer insights about the value and significance of the human body . . . the human body is created in the image of God and therefore there are limits on what human beings can do with their own bodies and those of others. Although several methods of transferring human tissues and cells are acceptable within the Jewish, Catholic, and Protestant traditions priority is given to explicit gifts.[3]

Enriched with the teachings of the past and led by a future vision of the body as inviolable and venerable, the body revolution offers a clear alternative to the engineering and marketing of life. To balance the technological obsession with efficiency, it offers the principle of empathy. In opposition to the attempt to extend market principles into the body, it offers an alternative mode of human interaction: the gift principle.

Empathy Over Efficiency

Our current behavior toward the body, as toward most of the natural world, is governed by the principle of efficiency. We see our bodies as biological machines that are to be used efficiently and effectively. Whether in labor or in medicine, the code is clear: Minimum input for maximum output in minimum time. This appears sensible. Who, after all, is opposed to efficiency?

Yet if we treated our children solely on the basis of efficiency (minimum food and affection for maximum loyalty and performance in school), our behavior would correctly be viewed as pathological. In the same way, if we began to treat our friends in an efficient manner, or even our pets, those close to us would recommend immediate psychotherapy. In daily life we treat nothing that we love or care about solely or even primarily on the basis of efficiency, but on the basis of empathy. Whether child, spouse, friend, or pet, love—not efficiency—is the primary mover. Yet this realization has not made its way into our public policies on the body, or toward the rest of the natural world.

In this book we have seen the unfortunate world created when efficiency is not balanced by empathy. We have seen that those adhering to the dogma of efficiency do often end by loving compulsion. We have witnessed that as each scientific and technological advance has given society more power over nature or the body, it has been translated into the power of some over those weaker or as yet unborn. The proponents of eugenics sterilized and destroyed the inefficient among us. Over the last decade, we have begun to treat the brain dead and their bodies as little more than containers of valuable body parts. We are changing the method and manner of abortion to more efficiently retrieve fetal parts, occasionally from fetuses that are still alive. We have also begun to sacrifice the dignity of human reproduction and the life of embryos in a search for more efficient reproduction. As we have seen, many Americans have said that they would abort a child if it is predisposed to obesity. We engineer human genes into the permanent genetic code, the germline, of animals to make them more efficient food sources or laboratory tools. Plans for germline intervention in the human genetic code are also underway. Despite setbacks, and the creation of mutants, scientists are cloning larger mammals in the search for a perfectly efficient breeding method. Many predict the cloning of humans is next.

Whatever the problems of the past, present, or future, many will still be suspicious of a reliance on empathy over efficiency and mechanism. Doesn't technological progress depend on efficiency and using our knowledge to harness, exploit, and control nature, including our bodies? Isn't that just the way it is? What about organ transplants? They are saving lives, and fetal transplants may also. Reproductive technologies are creating babies for couples who could not have them otherwise. And we might just be able to genetically engineer ourselves into more fit bodies that might live far longer.

That is what technology and the drive toward efficiency can produce. What can empathy do to compare with the accomplishments of mechanism?

Our modern consciousness is so rooted in efficiency and the principle of gaining power over nature that we find it difficult even to imagine an alternative way of thinking about the body. But instead of pursuing knowledge to enforce efficiency, power, and control, we might just as easily use our new biological knowledge to become a better partner with our own bodies and the rest of creation. The psychological and scientific results brought about by the revival and development of empathy as a scientific ethic could be formidable.

This empathetic principle, when applied to the body, would undo the dangerous mechanism propounded by La Mettrie over two centuries ago. Most of us can remember empathetic and intuitive lessons from our early science classes—lessons that undermine the mechanistic paradigm. For example, nothing seems more magical than the spontaneous regeneration of lost body parts. As children we were both amazed and envious at how starfish or earthworms could generate new parts for themselves when arms and other body sections were lost through accident or experiment. Early-grade biology classes often conducted experiments in which students cut a flatworm into pieces and watched as each piece—a head, a tail, a side, or a mere slice—grew into a complete worm. We humans regenerate our blood and our skin, and even our internal organs have some regenerative power—if part of our liver is lost, new liver tissue develops to replace it.[4]

Processes of regeneration suggest that organisms have an invisible cohesion that is more than the sum of their parts. Each segment of the flatworm or starfish seems to contain an implicit wholeness that goes beyond its material structure and guides each part into recreating the whole. The ability to regenerate thus appears to represent a significant challenge to the "biological machine" mind-set. As biochemist Rupert Sheldrake notes,

> One of the most striking ways in which living organisms differ from machines is their capacity to regenerate. No man-made

device has this capacity. If a computer, for example, is cut into pieces, these cannot, of course, form new computers. They remain pieces of a broken computer. The same goes for cars, telephones, and any other kind of machinery.[5]

Empathy with our bodies would also help crack the dualism that, since Descartes, has separated our bodies from our minds. Efficiency and scientism demand that we objectify our bodies as a first step in analyzing, controlling, and manipulating them. Yet the human form is in its own way a creator of the mind; mind and body cannot in reality be separated. Our bodily features closely correspond with basic aspects of our psyche. Take, for example, our upright posture. Anyone who sees an infant first struggle to stand feels instinctively the pride of our unique posture. Standing upright embodies basic human traits—awareness, awakeness, and a special kind of human will as we resist gravity. When the will is gone, so is the posture; we recline to rest or sleep. Moreover, when upright, our upper arms and hands are free, we no longer require them for support as do other mammals. This allows us the uniquely human activities of pointing, writing, gesturing, creating, and destroying. Our reach can exceed our grasp.

Moreover, other animals move in the direction of their digestive tracts; they move along the plane of their mouths, stomachs, and anuses. Humans, by contrast, stand upright; with feet on the ground and head uplifted, we travel not in the direction of digestion, but rather in the direction of our sight, our vision. Our posture opens us to the cosmos above. As psychologist Erwin Strauss notes, because of our "visionary" posture, "a galaxy and diluvium, the infinite and the eternal, enter the orbit of human interests."[6]

As we begin to transcend mechanism and dualism, we encounter a different view of the technologies of the human body shop. Empathetic consciousness about the body recognizes the limited value of the intrusive techniques of transplantation, reprotech, genetic screening, and genetic engineering, but does not idolize them. It notes that the majority of the patients for liver transplants are individuals who are suffering from liver disease due to alcohol consumption. Of the tens of thousands suffering from liver disease, transplants even at their maximum could save but a few hundred each year. The real solution to liver disease is not and never will be transplantation, but rather lies in prevention. We must change the stressful, efficiency-dominated work and life habits that lead to alcohol abuse and other compensatory addictions and develop those life habits that are more in tune with the rhythms our bodies need. The same is true of most heart disease. Heart transplants, again even at optimal levels, can barely make a dent in the millions suffering and dying from heart disease. We need to change our ways of working, living, and eating, which cause the vast majority of heart disease, and find daily routines that better fit our bodies.

Additionally, the "gee whiz" technologies of reproduction treat only an infinitesimal percentage of those millions suffering from infertility. And with the low percentage of success and high cost of reprotech, it is unlikely that the reproductive technology industry will be able to treat significantly more patients in the foreseeable future. The real causes of infertility are numerous body-abusive medical and personal practices. These include unnecessary hysterectomies, contraceptive devices that damage the body, and sexual behaviors that encourage promiscuity and lead to venereal disease. The preventive empathetic approach to infertility as opposed to the mechanistic reprotech approach would deal with the causes of infertility rather than further the use of intrusive and destructive technologies on women's and men's bodies. It would also seek to provide greater public and private support to reduce the tragic levels of infant mortality due to malnutrition, drug abuse, and lack of education. Saving these unborn should be given a far higher social priority than development of ever more exotic reproductive technologies.

The genetic screening and engineering of humans have brought us into a new eugenic age. Our increasing knowledge about genetics has been transformed into a number of intrusive technologies seeking to make the human body and its reproductive processes more efficient. However, our growing understanding of biology could lead to more sustainable and empathetic results. An empathetic approach teaches us that the wonderful diversity of human genes, including disease genes, is essential for our survival. As such, it is impossible to know which human genes are good and which are bad. Rather than eliminate human genes viewed as inefficient, we should celebrate our diversity. We should apply the same principle to other members of the living kingdom, using genetics to better understand and protect our dwindling species rather than to genetically engineer the natural world toward efficiency.

With empathy as our guide we are led to a new understanding of progress. For far too long the word "progress" has been used as an unquestioned rationale for the unlimited use of technology. Virtually any new invention that allows us to act or produce more efficiently, be it the car, nuclear power, or genetically engineered animals, is automatically viewed as progress. In modern times, progress has become synonymous with industrial and technological growth. Those opposing any new technology are routinely called "luddites" or "anti-progress." We have become so accepting of this use of the term that we have forgotten that progress, by itself, is an incomplete idea; to be coherent it should always be accompanied by the question "progress toward what?" We can only ascertain what progress is if we have a future vision of what we want. Only then will we know whether or not any particular technology helps us to progress toward that vision.

As we have described, the empathetic treatment of the body and the practices of the human body shop offer sharply contrasting visions of the future of the human person, two opposing views of what constitutes progress for the human body and other life-forms. The body shop future vision permits the sale of organs and fetal parts, subcontracts out having a baby, creates a breeder class to sell tissues, organs, and reproductive elements, and allows us to change the definition of life and death to suit the requirements of body parts demand. It also envisions a eugenic future in which the unfit are selected out before birth or genetically engineered after it. The empathetic body vision of the future focuses on a sacred image of the human form. It has an appreciation and awe for the diversity of all human and other life-forms. It places an emphasis on preventive medicine and more sustainable life-styles.

If we wish to realize the empathetic future vision rather than that of the body shop, the following specific biopolicies need to be implemented:

No expansion of the legal definition of death to include the higher brain dead. No use of cadavers or neomorts as storage

receptacles of organs. Respectful treatment and burial of the dead.

A moratorium on the use of induced-abortion fetuses for transplantation and research until the profound ethical and legal problems surrounding this practice are fully discussed and resolved. Issues such as consent, vivisection of fetuses, coercion to donate, changing the method and manner of abortion to harvest fetuses, and surreptitious payments for fetuses must be resolved. Additionally, we must never treat the fetus as merely the means to an end, no matter how altruistic those ends may be.

No eugenic use of "superior" sperm or eggs.

No cloning of human embryos, and maximum attempts to see that frozen embryos are given a chance at life.

Limits on the use of genetic screening of the unborn (amniocentesis, CVS, or preimplantation genetic screening of embryos) to ensure that screening is used only for detecting life-threatening disease. No use of prenatal screening to determine sex, weight, height, or other nondisease traits.

No genetic screening or monitoring of workers, and no discrimination against individuals in questions of employment or insurance or health coverage based on their genetic readout.

No use of genetically engineered drugs to alter or treat human traits that are the object of discrimination (height, pigmentation, and so on).

Limitation of gene therapy to the treatment of life-threatening disease. No use of gene engineering of humans for cosmetic or enhancement purposes.

A moratorium on the germline alteration and cloning of animals, including the engineering of human genes into animals, until there has been a full public debate on the issue and the ethical and environmental consequences of the genetic engineering of animals are better understood.

A ban on germline genetic therapy for the foreseeable future. We do not have the wisdom to know which genes are "good" and which genes are "bad."

A complete ban on the cloning of human beings.

The Gift Principle

The Trobriand Islanders of Melanesia are the subject of Bronislaw Malinowski's anthropological classic, *Argonauts of the Western Pacific*. The work, published in 1922, describes one group of the Massim peoples who occupy the South Sea islands. Malinowski spent several years living on these islands during World War I, primarily in the Trobriands, the northwesternmost group of the islands. He saw the Trobrianders as adventurous and hardy sailors. They reminded Malinowski of the mythological sailors who braved the perils of the Mediterranean Sea with Jason in the *Argosy*.

As we might expect from a people adept at traveling by sea, the Trobrianders were avid traders. Yet these West Pacific traders made a key distinction between various ways of dealing with the goods they exchanged. They dealt with some items through *Gimwali*, the straightforward trading of various products. Gimwali generally involved shrewd assessment of goods and tenacious bargaining. The Islanders had another means of exchange, however, called *Kula*. This involved a ceremonial exchange of gifts "carried out in a noble fashion, disinterestedly and modestly." The Kula gifts, primarily armshells and necklaces, move continually around a wide ring of islands in the Massim archipelago. Each moves in a circle. The red shell necklaces (considered to be male and worn by women) move clockwise around the islands; the armshells (viewed as female and worn by men) move counterclockwise. The revered Kula objects are not actually passed hand to hand but carried by canoe from island to island and from tribal partner to tribal partner, in journeys that often involve hundreds of miles. Malinowski remembers the gift-giving scene at one village:

> A native village on coral soil, and a small rickety platform temporarily erected under a pandanus thatch, surrounded by a number of brown, naked, men, and one of them showing me long, thin red strings, and big white, worn-out objects, clumsy to sight and greasy to touch. With reverence he also would name them and tell their history, and by whom and when they were worn, and how they changed hands, and how their temporary possession was a great sign of the importance and glory of the village.[7]

The Trobrianders strictly separated occasions for trade and occasions for gift giving. "The decorum of the Kula transaction is strictly kept, and highly valued. The natives distinguish it from barter, which they practice extensively. . . . Often when criticizing an incorrect, too hasty, or indecorous procedure of Kula, they will say: 'he conducts his Kula as if it were Gimwali [barter].'" Gimwali involves endless talk, but the Kula gift is given in silence.[8]

Malinowski's argonauts are not an isolated case. Gift giving has a long history of helping to organize vital elements of various societies' economic and social lives. In most cultures social cohesion itself relies on the strict separation between bartered items and those held to be above bartering. Cultures as diverse as the Native American Indians of the northwestern United States, the peasant social units of medieval England, the Samoans, and the ancient Romans distinguished between ordinary articles of consumption for sale and those articles not for sale. Those held outside the area of utilitarian trade are seen as *sacra*, sacred items—whether food, art, respected parts of nature, or religious objects—that are held to be venerable and priceless.

In many cultures sacra is associated with the major aspects of the human body's history. Gift-giving ceremonies involve fertility rites and threshold rites—welcoming in the newborn, and helping the dying cross the threshold into the next world. Virtually all cultures share the belief that the human body is a sacra, a venerable aspect of the human person, which by definition should not be sold. As noted by ethicist Thomas Murray, "Putting a price on the priceless . . . cheapens it."[9] To the extent that a body part partakes in the sense of life, it is sacra.

Clearly, the market system has eschewed the concept of sacra. It has invaded traditional areas of sacra, be they land, food, or the body, and made them fictitious commodities. It is long past due that we challenge the market hegemony over our society and begin balancing the market principle with the gift principle. We have much to learn from prior cultures, and the importance of gift giving and sacra may head the list.

Beginning to aggressively substitute the gift principle for that of the market will have many communal advantages. Sociologist Max Weber summed it up many decades ago:

The market community, as such, is the most impersonal relationship of practical life into which human beings can enter with one another. Where the market is allowed to follow its own tendencies, its participants do not look toward the person of each other, but only toward the commodity. There are no obligations of brotherliness or reverence, and none of those spontaneous human relations that are sustained by personal unions. They all would just obstruct the free development of the bare market relationship. Such absolute depersonalization is contrary to all elementary forms of human relations.[10]

Unlike the market, gift giving affirms a sense of community, charity, reverence, and a spontaneous sense of human relations. It is the antidote to the market system. As ethicist Thomas Murray explains:

Gifts create moral relationships that are more open-ended, less specifiable, and less contained than contracts. Contracts are well suited to the marketplace where a strictly limited relationship for a narrow purpose—trading goods or services—is desired. Gifts are better for initiating and sustaining more rounded human relationships, where future expectations are unknown, and where the exchange of goods is secondary in importance to the relationship itself.[11]

When the issue is human body parts, affirming the gift relationship is of even greater importance. The impersonal relations of the market not only demean the body, they demean those who buy and sell aspects of themselves. Throughout this book we have witnessed the appalling results of relegating the human body to the impersonal forces of the market and contracts. We have seen vampirism of the poor in the Third World for their blood and blood products. We have documented the extraordinary worldwide exploitation that results when the wealthy buy the organs of the poor. We have followed fetal procurers on their grim daily rounds, sorting through fetal remains to find those of value. We have heard the anguish of fathers who sold their sperm and now wish to have a relationship with their children, and the children who now search for their fathers with expectation and yet anger at having been the product of a sale. Surrogate mothers have found time and again that maternal bonds are far stronger than commercial bonds and have fought for years against heavy odds and media vilification to regain a relationship with their children. We have witnessed the unseemly sight of researchers and corporations fighting for patent rights over valuable genes that might be linked to intelligence or other desirable human traits. The lesson is clear. A sense of community and reverence are necessary if we are to keep a moral sense of our persons and our community and assure that the body remains a sacra.

In contrast to the horrors of the contracting and sale of human body parts, *giving* a body part creates moral cohesion in both the giver and the receiver. Note the responses of a number of blood donors who were asked by English economist Richard Titmuss about their reasons for donating (with spelling and punctuation preserved):

> Knowing I mite be saving somebody life.

> You cant get blood from supermarkets and chaine stores. People themselves must come forward.

> I thought it just a small way to help people—as a blind person other opportunities are limited.

Some saw giving as a way of saying thank-you for good health:

> Briefly because I have enjoyed good health all my life and in a small way this is a way of saying "Thank you" and a small donation to the less fortunate.

Others are giving back what was given to them:

> To try and repay in some small way some unknown person whose blood helped me to recover from two operations . . .

> Some unknown person gave blood to save my wifes life.[12]

Surveys of organ donors have revealed the same lack of self-interest, respect for the donated body part, and a deep feeling of reciprocity about body-parts gifts. In an increasingly commercialized medical market and a society that daily becomes more bureaucratized and impersonal, human body gifts affirm that we are still citizens together, that even anonymously we respect one another's needs and administer to one another's illnesses and disabilities. Human body gifts affirm that self-interest need not be our God. Catholic theologian Thomas Merton, writing of the Buddhist's begging bowl, describes the interdependence affirmed by giving. The begging bowl, he writes, "represents the ultimate theological root of the belief . . .in openness to the gifts of all beings as an expression of the interdependence of all beings."[13] It is the same with gifts of the body to help others: In so doing we reaffirm the dignity of our persons and our community.

To implement the gift principle over that of the market, we will need to promulgate the following specific biopolicies:

Continue to keep our system of blood donation for transfusions nonpaid, and close down the commercial sale of blood for pharmaceutical and research use.

Strongly support the United States and other countries in banning the sale of organs for transplantation, and extend the ban to include organs used for research.

Strongly support the ban on the sale of fetal parts and ensure that it is being strictly enforced.

Promulgate bans on the sale of sperm, eggs, and embryos.

Enact an international abolition of surrogate motherhood, and advocate criminal penalties for surrogate brokers.

Promulgate an international ban on the patenting of all life-forms, including genetically engineered animals and human cells, genes, embryos, organs, and other body parts.

———

The human body is the final battleground on which to fight the doctrines of mechanism and the market that have dominated our thinking for so long. We must balance the efficient and mechanistic approach with empathy toward our bodies and one another. We must balance the market ideology by establishing a gift principle for non-commodities, including sacra such as the human body. Personally and socially, we must rediscover a sense of the reverence and meaning of our bodies. We must finally understand that the human body shop injures not only its individual victims, but the very image of the sacred.

NOTES

Introduction

1. U.S. Congress, Office of Technology Assessment (OTA), *New Developments in Biotechnology—Background Paper: Public Perceptions of Biotechnology, OTA-BP-BA-45* (Washington, D.C.: U.S. Government Printing Office, May 1987), 3.

CHAPTER 1

Blood Tithes

1. Richard M. Titmuss, *The Gift Relationship: From Human Blood to Social Policy* (New York: Pantheon Books, 1971), 171, 172, n.5.

2. Titmuss, *The Gift Relationship*, 15.

3. Matthew 26:27–28.

4. P. J. Hagen, *Blood: Gift or Merchandise* (New York: Alan R. Liss, 1982), 11; also Titmuss, *The Gift Relationship*, 17, n.1.

5. Cited in Charles Singer, *A Short History of Anatomy and Physiology from the Greeks to Harvey* (New York: Dover, 1957), 182.

6. David F. Channell, *The Vital Machine* (New York, Oxford: Oxford University Press, 1991), 33–34.

7. Cited in Earle Hackett, *Blood, the Biology, Pathology, and Mythology of the Body's Most Important Fluid* (New York: Saturday Review Press, 1973), 183.

8. Hagen, *Blood*, 11.

9. Cited in P. L. Mollison, *Blood Transfusion in Clinical Medicine*, 2d ed. (London: Blackwell, 1956), 23.

10. Cited in W. L. Palmer, "Serum Hepatitis Consequent to Transfusion of Blood," *Journal of the American Medical Association* 180, no. 13 (1962): 1123.

11. Titmuss, *The Gift Relationship*, 18.

12. Hackett, *Blood*, 45.

13. Hackett, *Blood*, 45.

14. Hackett, *Blood*, 151–54.

15. Hackett, *Blood*, 155.

16. In the Matter of Community Blood Bank of the Kansas City Area, Inc., et al., docket 8519 (1966), 735.

17. Titmuss, *The Gift Relationship*, 158.

18. Titmuss, *The Gift Relationship*, 159, 160.

19. Community Blood Bank of the Kansas City Area, 735–39, 770, 771, 839; Titmuss, *The Gift Relationship*, 159, 160.

20. Community Blood Bank of the Kansas City Area, 739.

21. Community Blood Bank of the Kansas City Area, 838.

22. Community Blood Bank of the Kansas City Area, 838, 839.

23. Titmuss, *The Gift Relationship*, 161, 162.

24. Community Blood Bank of the Kansas City Area, 901.

25. Community Blood Bank of the Kansas City Area, 904.

26. Titmuss, *The Gift Relationship*, 161, 162.

27. Community Blood Bank of the Kansas City Area, Inc., v. F.T.C., 405 F.2d 1011, 1012, 1022 (1969).

28. Hagen, *Blood*, 168-9.

29. Hagen, *Blood*, 169.

30. Hagen, *Blood*, 168-9.

31. Hagen, *Blood*, 1.

32. Titmuss, *The Gift Relationship*, 238, 239.

33. Robin Herman, "Continuing Vigilance Over the Blood Supply," *Washington Post* (April 21, 1992): Health 8, 9.

34. Douglas MacN. Surgenor, Ph.D., Edward L. Wallace, Ph.D., Steven H. S. Hao, M.B.A., Richard H. Chapman, B.A., "Collection and Transfusion of Blood in the United States, 1982-1988," *New England Journal of Medicine* (June 7, 1990): 1646-1648.

35. Forum, "Sacred or For Sale?" *Harper's* 281, no. 1685 (October 1990): 39.

36. "Sacred or For Sale?" 39.

37. "Sacred or For Sale?" 49.

38. Blumberg, B. S., I. Millman, W. T. London, et al., "Ted Slavin's Blood and the Development of HBV Vaccine," Letter to the Editor, *New England Journal of Medicine* 312, no. 3 (1985): 189.

39. Letter from Robert Reilly, President, American Blood Resources Association (ABRA), to Katherine Matthews, Foundation on Economic Trends, April 27, 1992, ABRA Reference No. A10685; Robert Reilly, "The International Plasma Market," in James L. MacPherson and Lori B. Beaston, eds., *Adequacy of the Nation's Blood Supply* (Washington, D.C.: Council of Community Blood Centers, 1990), 89; Hagen, *Blood*, 11, 168.

40. T. C. Drees, cited in Hagen, *Blood*, 66.

41. Margaret Cramer Green v. Commissioner of Internal Revenue Service, 74 T.V. 1229 (1980).

42. Green v. Commissioner of Internal Revenue Service.

43. Green v. Commissioner of Internal Revenue Service.

CHAPTER 2

Transplanting Profits

1. Pakrash Chandra, "Kidneys for Sale," *World Press Review* 38 (February 1991): 53.

2. Quoted in Russell Scott, *The Body as Property* (New York: Viking Press, 1981), 18.

3. Quoted in Felix T. Rapaport and Jean Dausset, *Human Transplantation* (New York and London: Grune & Stratton, 1968), 4, 5.

4. Russell Scott, *The Body as Property* (New York: Viking Press, 1981), 19. See also Nancy Rosenberg, *New Parts for People* (New York: Norton, 1969), 2–4.

5. Rosenberg, *New Parts,* 61; Scott, *The Body as Property,* 19, 20.

6. Rosenberg, *New Parts,* 53–58; David Lamb, *Organ Transplants and Ethics* (London and New York: Routledge, 1990), 12, 13.

7. Source United Network for Organ Sharing (1100 Boulder Parkway, Suite 500, P.O. Box 13770, Richmond, VA 23225-8770), preliminary statistics for 1991.

8. "Re-usable Body Parts," *Washington Post* (May 28, 1991): Health.

9. "Re-usable Body Parts."

10. "The New Era of Transplants," *Newsweek* (August 29, 1983): 40.

11. Kim Painter, "Does the End Justify Paying for Organs?" *USA Today* (September 24, 1991): Life 1D; See also, "Be Open to New Ways to Spur Organ Donations," *USA Today* (September 24, 1991): 10A.

12. Lamb, *Organ Transplants,* 134.

13. Scott, *The Body as Property,* 1.

14. See The National Organ Transplant Act (NOTA), 42 U.S.C. 274 (1984).

15. Lamb, *Organ Transplants,* 134.

16. See NOTA, 42 U.S.C. 274.

17. Lamb, *Organ Transplants,* 138

18. "Trading Flesh Around the Globe," *Time* 137 (June 17, 1991): 61.

19. "Trading Flesh Around the Globe," 61.

20. Chris Hedges, "Egypt's Desperate Trade: Body Parts for Sale," *New York Times* (September 23, 1991): 1.

21. "Trading Flesh Around the Globe," 61.

22. Chandra, "Kidneys for Sale," 53.

23. *Parade,* Intelligence Report (September 13, 1992): 14.

24. Hedges, "Egypt's Desperate Trade," 1.

25. Quoted in Lamb, *Organ Transplants,* 134, 135.

26. "Be Open to New Ways to Spur Organ Donations," 10A.

27. UNOS Update, November 1992, Volume 8, Issue 11, p. 13.

28. Lori Andrews, Jack Kevorkian, Andrew Kimbrell, William May, "Sacred or for Sale? The Human Body in the Age of Biotechnology," *Harper's* (October 1990): 50.

29. Andrews, *et al.,* "Sacred or for Sale?" 49.

30. Lamb, *Organ Transplants,* 135.

31. UNOS Update, November 1992, Volume 8, Issue 11, p. 14.

32. Rita L. Marker, "Don't Pay for Organs," *USA Today* (September 24, 1991): 10A.

33. Andrews, *et al.,* "Sacred or for Sale?" 50.

34. See "Harvard Criteria for Irreversible Coma," *Medical World News* (April 28, 1986): 86.

35. Quoted in Susan Jenks, "The Lingering Image of Life," *Medical World News* (April 28, 1986): 90.

36. Jenks, "The Lingering Image of Life," 91.

37. Jenks, "The Lingering Image of Life," 91.

38. Joseph M. Darby, M.D.; Keith Stein, M.D., Ake Grenvik, M.D., Ph.D.; Susan A. Stuart, R.N., "Approach to Management of the Heartbeating 'Brain Dead' Organ Donor," *Journal of the American Medical Association* 261, no. 15 (April 21, 1989): 2222.

39. Stuart J. Younger, M.D.; Seth Landefeld, M.D.; Claudia J. Coulton, Ph.D.; Barbara W. Juknialis, M.A.; Mark Leary, M.D., "'Brain Death' and Organ Retrieval; A Cross-sectional Survey of Knowledge and Concepts Among Health Professionals," *Journal of the American Medical Association* 261, no. 15 (April 21, 1989): 2205.

40. Jay A. Friedman, "Taking the Camel by the Nose: The Anencephalic as a Source for Pediatric Organ Transplants," *Columbia Law Review* 90 (May 1990): 917, 921, n.18.

41. "Infant in Organ Dispute Dies," *Newsday* (March 31, 1992): 7.

42. Quoted in Charles Krauthammer, "The Case of Baby Theresa," *Washington Post* (April 3, 1992): 28.

43. Friedman, "Taking the Camel by the Nose," 977.

44. Krauthammer, "The Case of Baby Theresa."

45. Darby, *et al.*, "Approach to Management of the Heartbeating 'Brain Dead' Organ Donor," 2222.

46. Lamb, *Organ Transplants*, 44.

47. Lamb, *Organ Transplants*, 55, 56, 57.

48. Lamb, *Organ Transplants*, 24.

49. Lamb, *Organ Transplants*, 65.

50. Quoted in Lamb, *Organ Transplants*, 57.

51. Cited in Richard John Neuhaus, ed., *Guaranteeing the Good Life: Medicine and the Return of Eugenics* (Grand Rapids, MI: Wm. B. Eerdmans, 1990), 11.

52. Quoted in Kathleen Stein, "Last Rights," *Omni* (September 1987): 67.

53. Quoted in Lamb, *Organ Transplants*, 51, 52.

CHAPTER 3

Harvesting the Unborn

1. George Archibald, "Embryonic Enterprises," *Washington Times* (January 6, 1992): A1, A7; Dorothy Vawter, Warren Kearney, Karen G. Gervais, Arthur L. Caplan, Daniel Garry, Carol Tauer, "The Use of Human Fetal Tissue: Scientific, Ethical, and Policy Concerns; A Report of Phase I of an Interdisciplinary Research Project Conducted by the Center for Biomedical Ethics" (Minneapolis: University of Minnesota, January 1990), 194 ; personal interview with George Archibald, April 14, 1992.

2. Archibald, "Embryonic Enterprises," A1, A7; Dorothy Vawter, et al., "The Use of Human Fetal Tissue," 194 ; personal interview with George Archibald, April 14, 1992.

3. Archibald, "Embryonic Enterprises," A1, A7.

4. David H. Smith, et al., "Using Human Fetal Tissue for Transplantation and Research: Selected Issues," paper submitted to the Office for Protection from Research Risks, National Institutes of Health, in consultation with the Executive committee of the Human Fetal Tissue Transplantation Research Panel (September 1, 1988), 13; cited in Vawter, et al., "The Use of Human Fetal Tissue," 191.

5. Archibald, "Embryonic Enterprises," A7.

6. Archibald, "Embryonic Enterprises," A7; Vawter, et al., "The Use of Human Fetal Tissue," 195.

7. Timothy J. McNulty, "Murky Moral Issues Surround Fetal Research," *Chicago Tribune* (July 27, 1987).

8. Emmanuel Thorne, "Trade in Human Tissue Needs Regulation," *Wall Street Journal* (August 19, 1987).

9. Vawter, et al., "The Use of Human Fetal Tissues," 11, B1.

10. Gina Kolata, "Miracle or Menace," *Redbook* (September 1990), 175; Larry Thompson, "American Undergoes Fetal Cell Implant," *Washington Post* (November 15, 1988): Health 5.

11. Sarah Glazer, "The Fight Over Fetal Tissue," *Washington Post* (June 30, 1992): Health 11.

12. Georgina Ferry, "New Cells for Old Brains," *New Scientist* (March 24, 1988): 54–57.

13. Vawter, et al., "The Use of Human Fetal Tissue," 67, 106, 110.

14. Glazer, "The Fight Over Fetal Tissue," 11.

15. Antonin Scommenga, quoted in McNulty, "Murky Moral Issues."

16. "Fetuses Could Be Conceived to Be Aborted for Transplants, Scientists Say," Reuters Wire Service (November 20, 1987).

17. "Fetuses Could Be Conceived."

18. "Medical Technology: Handle with Care," *USA Today* (June 4, 1991): 10A.

19. Gina Kolata, "More Babies Being Born to Be Donors of Tissue," *New York Times* (June 4, 1991): C3.

20. Kolata, "More Babies Being Born."

21. Sandra Blakeslee, "New Medical Research Tool: Human Tissues in Lab Mice," *New York Times* (October 30, 1990): A1.

22. Blakeslee, "New Medical Research Tool," C3; Andrew Tyler, "Mice as People," *The Independent Magazine* (April 20, 1991): 29.

23. Blakeslee, "New Medical Research Tool," C3; Tyler, "Mice as People," 29.

24. Blakeslee, "New Medical Research Tool," C3; Tyler, "Mice As People," 26.

25. McNulty, "Murky Moral Issues."

26. Henry A. Waxman, "Research That Could Save Lives," *Washington Post* (May 21, 1991): A21.

27. Stephen G. Post, "Fetal Tissue Transplant: The Right to Question Progress," *America* (January 12, 1991): 14.

28. Peggy Finston, M.D., with Gregory J. Millman, "Alone and Undaunted by the Issues . . ." *OTC Review* (October 1989): 26; Sabin Russell, "Biotech Firm Drops Plan for Use of Fetal Tissue," *San Francisco Chronicle* (December 14, 1990): Business C1, C4.

29. Finston and Millman, "Alone and Undaunted by the Issues," 26.

30. Petition from the Foundation on Economic Trends to the Honorable Otis R. Bowen, September 14, 1987 (on file with the Foundation on Economic Trends, Washington, D.C.).

31. Public Law 100-607, Title IV—Organ Transplant Amendments of 1988, Sec. 407. Fetal Organ Transplants.

32. Russell, "Biotech Firm Drops Plan for Use of Fetal Tissue," C1, C4.

33. Kolata, "Miracle or Menace," 176, 216.

34. Vawter, et al., "The Use of Human Fetal Tissue," 169.

35. Carlyn Gerster, M.D., "Fetal Transplant—A Critique," *APDA Newsletter* (September 1991): 1.

36. Vawter, et al., "The Use of Human Fetal Tissue," 112, 113.

37. Kolata, "Miracle or Menace," 176.

38. Vawter, et al., "The Use of Human Fetal Tissue," 161.

39. Keith A. Crutcher, Ph.D., "In Support of the NIH Moratorium on the Use of Federal Funds for the Therapeutic Use of Human Fetal Tissue Derived from Elective Abortion," unpublished (April 2, 1990).

40. Petition from the Foundation on Economic Trends (Foundation) to James B. Wyngarden, Director National Institutes of Health (NIH), February 26, 1987; NIH response to the Foundation, September 25, 1987; Foundation response to NIH, February 26, 1988 (on file with the Foundation on Economic Trends, Washington, D.C.). Note: The author has been policy director and attorney for the Foundation of Economic Trends since 1986.

41. Vawter, et al., "The Use of Human Fetal Tissue," 162.

42. Vawter, et al., "The Use of Human Fetal Tissue," 162.

43. Crutcher, "In Support of the NIH Moratorium," 12.

44. Vawter, et al., "The Use of Human Fetal Tissues," 163.

45. See Vawter, et al., "The Use of Human Fetal Tissues," Appendix C.

46. Vawter, et al., "The Use of Human Fetal Tissue," 196.

47. Vawter, et al., "The Use of Human Fetal Tissue," 162.

48. Quoted in Dave Andrusko, "Another Look at Fetal Transplants," *Right to Life News* (March 26, 1991): 16.

49. David Lamb, *Organ Transplants and Ethics* (London and New York: Routledge, 1990), 70.

50. Quoted in Andrusko, "Another Look at Fetal Transplants," 16.

51. James Christie, "Fetal Tissue May Be Next Doping Issue," *The Washington Times* (August 22, 1992).

52. "Outraging Public Decency with Foetus Earrings," Law Report, *The Times* (London) (July 12, 1990): 37.

53. "Fetal Tissue Right or Wrong?" *Redbook* (December 1990): 170.

54. Vawter, et al., "The Use of Human Fetal Tissue," 177.

55. Vawter, et al., "The Use of Human Fetal Tissue," 178–82.

56. George Archibald, "NIH Skirts Ban on Transplants of Fetal Tissue," *Washington Times* (January 6, 1992): A7.

CHAPTER 4

The Business of Baby-Making

1. Quoted in Linda Campbell, "Advances with Embryos Leave Laws Far Behind," *Chicago Tribune* (April 8, 1992): C1.

2. Glenn Kramon, "The Infertility Chain: The Good and Bad in Medicine," *New York Times* (June 19, 1992): D1.

3. William D. Mosher, Ph.D., and William F. Pratt, Ph.D., "Fecundity and Infertility in the United States, 1965–88," advance data, from *Vital Health Statistics of the National Center for Health Statistics* 192 (December 4, 1990): 1.

4. Mosher and Pratt, "Fecundity and Infertility," 1.

5. Mosher and Pratt, "Fecundity and Infertility," 1.

6. U.S. Congress, Office of Technology Assessment, *Infertility: Medical and Social Choices, OTA-BA-358* (Washington, D.C.: U.S. Government Printing Office, May 1988), 61–71.

7. Annette Baran and Reuben Pannor, *Lethal Secrets, the Shocking Consequences and Unsolved Problems of Artificial Insemination* (New York: Warner Books, 1989), 25.

8. OTA, *Infertility: Medical and Social Choices,* 61–71; Baran and Pannor, *Lethal Secrets,* 25.

9. Mosher and Pratt, "Fecundity and Infertility," 1.

10. OTA, *Infertility: Medical and Social Choices,* 148, 160.

11. See Barbara Katz Rothman, "The Frightening Future of Baby-Making," *Glamour* (June 1992): 211.

12. Alison Leigh Cowan, "Can a Baby-Making Venture Deliver," *New York Times* (June 1, 1992): D1, D6.

13. Cowan, "Can a Baby-Making Venture Deliver," D1, D6.

14. Mosher and Pratt, "Fecundity and Infertility," 5.

15. Mosher and Pratt, "Fecundity and Infertility," 6.

16. OTA, *Infertility: Medical and Social Choices,* 53.

17. Quoted in Ellen Hopkins, "Tales from the Baby Factory," *New York Times* (March 15, 1992): Section 6.

18. Cowan, "Can a Baby-Making Venture Deliver," D1.

19. "Fertility Procedures' Track Records," *Washington Post* (January 21, 1992): Health 5.

20. Hopkins, "Tales from the Baby Factory."

21. Hopkins, "Tales from the Baby Factory."

22. Hopkins, "Tales from the Baby Factory."

23. Paul Taylor, "U.S. Having Mixed Success in Cutting Infant Mortality," *Washington Post* (March 27, 1992): A16.

CHAPTER 5

The Seeds of Life

1. Annette Baran and Reuben Pannor, *Lethal Secrets, the Shocking Consequences and Unsolved Problems of Artificial Insemination* (New York: Warner Books, 1989), 97.

2. Wilfred J. Finegold, M.D., *Artificial Insemination*, 2d ed. (Springfield, IL: Charles C. Thomas, 1976), 6.

3. Finegold, *Artificial Insemination*, 7.

4. U.S. Congress, Office of Technology Assessment, *Artificial Insemination: Practice in the United States Summary of a 1987 Survey-Background Paper, OTA-BP-BA-48* (Washington, D.C.: U.S. Government Printing Office, August 1988), 40.

5. OTA, *Artificial Insemination*, 67.

6. OTA, *Artificial Insemination*, 41.

7. OTA, *Artificial Insemination*, 3, 4, 15, 16.

8. OTA, *Artificial Insemination*, 48, 49.

9. OTA, *Artificial Insemination*, 32, 33.

10. Baran and Pannor, *Lethal Secrets*, 86.

11. Baran and Pannor, *Lethal Secrets*, 98.

12. Robin Herman, "When the 'Father' Is a Sperm Donor," *Washington Post* (February 11, 1992): Health, 10.

13. Herman, "When the 'Father' Is a Sperm Donor," 10.

14. Bud Norman, "Romance Cut Short by Family Ties," *San Jose Mercury-News* (May 1988): A1.

15. Herman, "When the 'Father' Is a Sperm Donor," 12.

16. Robin Herman, "An Unconventional Option, a Delightful Baby," *Washington Post* (February 11, 1992): Health 14.

17. Herman, "When the 'Father' Is a Sperm Donor," 12.

18. Baran and Pannor, *Lethal Secrets*, 37, 38.

19. Herman, "When the 'Father' Is a Sperm Donor," 12.

20. Rochelle Sharpe, "Love Can't Take Away Pain of Sperm Donor's Offspring," *El Paso Times* (December 21, 1986): 24A.

21. Sharpe, "Love Can't Take Away Pain," 24A.

22. Sharpe, "Love Can't Take Away Pain," 24A.

23. Sharpe, "Love Can't Take Away Pain," 24A.

24. Herman, "When the 'Father' Is a Sperm Donor," 13.

25. Sally Squires, "Shopping for Safe Sperm," *Washington Post* (February 11, 1992): Health 12.

26. Squires, "Shopping for Safe Sperm," 11.

27. Herman, "When the 'Father' Is a Sperm Donor," 13.

28. U.S. Congress, Office of Technology Assessment, *Infertility: Medical and Social Choices, OTA-BA-358* (Washington, D.C.: U.S. Government Printing Office, May 1988), 329–45.

29. Herman, "When the Father Is a Sperm Donor," 12.

CHAPTER 6
The Price of Eggs

1. Gina Kolata, "Young Women Offer to Sell Their Eggs to Infertile Couples," *New York Times* (Novermber 10, 1991).

2. See Robert P. S. Jansen, "Sperm and Ova as Property," *Journal of Medical Ethics* 11 (1985): 123.

3. Jansen, "Sperm and Ova as Property," 128.

4. Gina Kolata, "Young Women."

5. Robin Herman, "Egg Donation Centers More Accessible Today," *Washington Post* (July 14, 1992): Health, 5.

6. Kolata, "Young Women Offer to Sell Their Eggs."

7. Kolata, "Young Women Offer to Sell Their Eggs."

8. Joan Liebmann-Smith, "Medical Miracle . . . or Baby Selling?" *Redbook* (November 1992):124–26.

9. Liebmann-Smith, "Medical Miracle," 124.

10. Sally Squires, "The New Motherhood," *Washington Post* (February 12, 1991): Health, 15.

11. Liebmann-Smith, "Medical Miracle," 124.

12. Molly Gordy, "Egg Donors Offer the Gift of Life," *Newsday* (April 20, 1992): 21.

13. Kolata, "Young Women Offer to Sell Their Eggs."

14. Kolata, "Young Women Offer to Sell Their Eggs."

15. Squires, "The New Motherhood," 15.

16. Gordy, "Egg Donor Rejected," 6.

17. Gordy, "Egg Donor Rejected," 6.
18. Gordy, "Egg Donor Rejected," 6.
19. Gordy, "Egg Donor Rejected," 6.
20. Liebmann-Smith, "Medical Miracle," 120.
21. OTA, *Infertility: Medical and Social Issues,* 329–55.

CHAPTER 7

Embryo Imbroglio

1. Judy Licht, "Frozen in Time, Storing of Embryos Boosts the Chances of Pregnancy—and Raises Ethical Questions," *Washington Post* (November 26, 1991): Health 12.
2. Ronald Kotulak and Peter Gorner, "Babies by Design," *Chicago Tribune* (March 3, 1991): C14.
3. U.S. Congress, Office of Technology Assessment, *Infertility: Medical and Social Choices,* OTA-BA-358 (Washington, D.C.: U.S. Government Printing Office, May 1988), 298.
4. Judy Licht, "Frozen in Time," *Washington Post* (November 26, 1991): Health, 10.
5. Licht, "Frozen in Time," 10.
6. OTA, *Infertility: Medical and Social Choices,* 252.
7. OTA, *Infertility: Medical and Social Choices,* 252.
8. OTA, *Infertility: Medical and Social Choices,* 252.
9. OTA, *Infertility: Medical and Social Choices,* 259.
10. Linda F. Campbell, "Advances with Embryos Leave Laws Far Behind," *Chicago Tribune* (April 8, 1990): C1.
11. Roan, "Ethics and the Science of Birth," A1.
12. OTA, *Infertility: Medical and Social Choices,* 355.
13. Roan, "Ethics and the Science of Birth," A1.
14. Campbell, "Advances with Embryos," C1.
15. OTA, *Infertility: Medical and Social Choices,* 208–9.
16. See Davis v. Davis v. King, in the Circuit Court for Blount County, Tennessee, at Maryville, docket E-14496, September 21, 1989, 3–5.
17. Davis v. Davis v. King, 3–5.
18. Davis v. Davis v. King, 3–5, B3.
19. Davis v. Davis v. King, 3–5, B12, B13.
20. Davis v. Davis v. King, 3–5, 6, 7.
21. Davis v. Davis v. King, 3–5, B4, B5, B13.
22. Davis v. Davis v. King, 3–5, B18.
23. Davis v. Davis v. King, 3–5.
24. Davis v. Davis v. King, 3–5, 2.

25. Mark Curriden, "Tennessee Embryos Case Becomes Legal Nightmare," *Dallas News* (July 29, 1990): A1.

26. Curriden, "Tennessee Embryos Case," A1.

27. Curriden, "Tennessee Embryos Case," A1.

28. Curriden, "Tennessee Embryos Case," A1.

29. Curriden, "Tennessee Embryos Case," A1.

30. See Davis v. Davis, Court of Appeals of Tennessee, No. 180, September 13, 1990.

31. See Davis v. Davis, in the Supreme Court of Tennessee, at Knoxville, No. 34, June 1, 1992, 4.

32. See Davis v. Davis, Supreme Court of Tennessee, 39.

33. See Davis v. Davis, Supreme Court of Tennessee, 21.

34. See York v. Jones, 717 F. Supp. 421 (E. Va. 1989).

CHAPTER 8
Baby-Selling, Pure and Simple

1. *USA Today* (September 2, 1988), 5D.

2. 1990 figures provided by the National Coalition Against Surrogacy, Washington, D.C.

3. Rebecca Powers and Sheila Gruber Belloli, "The Baby Business," part 4, *Detroit News* (September 20, 1989).

4. *The Business of Surrogate Parenting* (Albany, NY: New York State Department of Health, April 1992), 3.

5. *The Business of Surrogate Parenting*, 3; also Rebecca Powers and Sheila Gruber Belloli, "The Baby Business," part 1, *Detroit News* (September 17, 1989): 1.

6. See, for example, "Surrogate Parenting Agreement" between Mary Beth White-head, Richard Whitehead, and William Stern, recital 5 (on file with author).

7. "Surrogate Parenting Agreement" between Mary Beth Whitehead, Richard Whitehead, and William Stern, recital 10.

8. Quoted in Susan Edmiston, "Whose Child Is This?" *Glamour* (November 1991): 237.

9. *The Business of Surrogate Parenting*, 3.

10. Powers and Belloli, "The Baby Business," part 2, *Detroit News* (September 18, 1989): C1.

11. Powers and Belloli, "The Baby Business," part 1, "Shattered Dreams," *Detroit News* (September 17, 1989).

12. Powers and Belloli, "The Baby Business," part 2.

13. Powers and Belloli, "The Baby Business," part 2.

14. Powers and Belloli, "The Baby Business," part 2.

15. Powers and Belloli, "The Baby Business," part 2; David Grogan and Beth Austin, "Little Girl, Big Trouble," *People* (February 20, 1989): 37–40.

16. Powers and Belloli, "The Baby Business," part 2.

17. Tamar Lewin, "Custody Case in Ohio Ends in Slaying and Prison Term," *New York Times* (December 8, 1990).

18. Powers and Belloli, "The Baby Business," part 2.

19. "Abuses in Surrogacy," in information packet from the National Coalition Against Surrogacy, Washington, D.C.

20. Rebecca Powers and Sheila Gruber, "Baby Broker Pleads Guilty to Mail Fraud," *Detroit News* (May 12, 1992): 1A.

21. Claudia Levy, "Md. Surrogate Parenting Agency Puts Business on Hold," *Washington Post* (March 31, 1988): C1; letter from Philip and Cynthia Custer to Maryland Attorney General Khanna, March 11, 1988; letter from Steven Litz of Surrogate Mothers, Inc., to Harriet Blankfield, Infertility Associate Inc., March 25, 1988; Claudia Levy, "Surrogate Mom Gives Up Baby After Dispute," *Washington Post* (April 23, 1988): G1.

22. Powers and Belloli, "The Baby Business," part 2.

23. Rochelle Sharpe, "Suit Challenges Constitutionality of Surrogate Parenting," *Michigan State Journal* (October 3, 1986): 1B.

24. Rochelle Sharpe, Gannett News Service, "Surrogate Mother: Ethical Unknown," copyright 1986.

25. Powers and Belloli, "Shattered Dreams," 6s.

26. Sharpe, "Surrogate Mother: Ethical Unknown."

27. "The Business of Surrogate Parenting," 4.

28. See, for example, Kathleen A. Welsh, "Not Just Anyone's Baby," unpublished (on file with the National Coalition Against Surrogacy).

29. U.S. Congress, Office of Technology Assessment, *Infertility: Medical and Social Choices, OTA-BA-358* (Washington, D.C.: U.S. Government Printing Office, May 1988), 269.

30. OTA, *Infertility: Medical and Social Choices*, 269.

31. OTA, *Infertility: Medical and Social Choices*, 269.

32. OTA, *Infertility: Medical and Social Choices*, 273–74.

33. "The Business of Surrogate Parenting," 4, fn. 2.

34. "Abuses in Surrogacy," National Coalition Against Surrogacy, Washington, D.C.

35. Contract between Noel Keane, Patty Nowakowski, and anonymous, dated July 27, 1987 (on file at the National Coalition Against Surrogacy).

36. Patty Nowakowski, "How Could I Let Them Separate My Twins?" *Redbook* (July 1990), 38.

37. Nowakowski, "How Could I Let Them," 40.

38. Nowakowski, "How Could I Let Them," 40.

39. Nowakowski, "How Could I Let Them," 41.

40. Statement of Patricia Foster, Press Conference National Coalition Against Surrogacy, August 31, 1987, Mayflower Hotel, Washington, D.C.

41. Anderson and Holland, "Caretaker Mum. A Baby She Doesn't Want to Surrender," *Sydney Sun-Herald* (July 22, 1984).

42. Statement of Mary Beth Whitehead (see footnote 40).

43. Nowakowski contract, recital 1.

44. For a psychiatric rebuttal of the screening argument, see (child psychiatrist) Robert E. Gould, "And What About Baby M's Ruined Life?" *New York Times* (March 26, 1987): A27.

45. Dr. Philip Parker, "The Psychology of the Pregnant Surrogate Mother: A Newly Updated Report of a Longitudinal Pilot Study," *American Orthopsychiatric Association Meeting,* Toronto, 1984, (April 9): 26.

46. Sobel, "Surrogate Mothers: Why Women Volunteer," *New York Times* (June 12, 1981).

47. Robin Winkler and Margaret Van Keppel, *Relinquishing Mothers in Adoption. Their Long Term Adjustment* (Melbourne: Institute of Family Studies, 1984).

48. "Abuses in Surrogacy," National Coalition Against Surrogacy, Washington, D.C.

49. R. Miller, "Surrogate Parenting: An Infant Industry Presents Society with Legal, Ethical Questions," *Obstetrics and Gynecological News* 3 (1983): 18.

50. See Dr. Philip Parker, "Motivation of Surrogate Mothers," *American Journal of Psychiatry* 140 (1983): 117–18.

51. L. Waller, "Born for Another," *Monash University Law Review* 10 (1984): 113, 119.

52. Gena Corea, *The Mother Machine* (New York: Harper & Row, 1985), 245.

53. "The Business of Surrogate Parenting," 12.

54. William M. Landes and Richard A. Posner, "The Economics of the Baby Shortage," *Journal of Legal Studies* 7: 323, 342–43.

55. Katha Pollitt, "The Strange Case of Baby M," *The Nation* (May 23, 1987).

56. See, for example, H. R. 2433; see also "Anti-Surrogacy Bill Passes in Nebraska, American Bar Association Panel Backs Curbs on Surrogate Motherhood," *New York Times* (February 6, 1988).

57. "The Business of Surrogate Parenting," 11.

58. In the Matter of Baby M (A-39-87), Supreme Court, State of New Jersey, Decided (February 3, 1988); Yates v. Keane and Hubers, file # 9758, 9772) (1, page 5), Circuit County of Gratiot, Michigan (January 25, 1988); In re Paul, 146 Misc. 2d 379, 550 N.Y.S. 2d 815 (Fam. Ct., Kings Co. 1990); Anonymous v. Anonymous, N.Y.L.J., October 10, 1991, at 27 (Fam. Ct., Bronx Co.).

59. Cited in L. Waller, "Born for Another," *Monash University Law Review* 10 (1984): 119.

60. See Petition for Review, Anna Johnson v. Mark Calvert and Crispina Calvert, Appeal from the Superior Court of Orange County, In the Supreme Court of California, at 11-13; Anna Johnson v. Mark Calvert, Daily Appellate Report, 12433-34.

61. Anna Johnson v. Mark and Crispina Calvert, Superior Court of the State of California for the County of Orange, No. x 63 31 90 consolidated with No. AD 57638, Honorable Richard N. Parslow, Jr., Judge Presiding Reporter's Transcript, October 22, 1990, at 2,4.

62. Parslow, 5.

63. Seth Mydans, "Science and the Courts Take a New Look at Motherhood," *New York Times* (November 4, 1990).

64. Michelle Harrison, M.D., "Anna Johnson Is a Mother, Not Just an Incubator," *Chicago Tribune* (October 8, 1990): Section 1, 13.

65. Mydans, "Science and the Courts."

66. Anna Johnson v. Mark Calvert et al, Supreme Court of California, SO23721, 21, 26, 27, 33.

67. Supreme Court, Order List, October 4, 1993, No. 93–5065.

68. Telephone interview between author and Richard Gilbert, November 22, 1991.

CHAPTER 9
The Perfect Baby

1. Quoted in Malcolm Gladwell, "Gene-Defect Tests Planned," *Washington Post* (December 31, 1990): A5.

2. Larry Thompson, "Cell Test Before Implant Helps Insure Healthy 'Test-Tube' Baby," *Washington Post* (April 27, 1992): A1.

3. Thompson, "Cell Test Before Implant," A1.

4. Thompson, "Cell Test Before Implant," A1.

5. Gladwell, "Gene-Defect Tests Planned," A1, A5.

6. Thompson, "Cell Test Before Implant," A5.

7. Ronald Kotulak and Peter Gorner, "Babies by Design," *Chicago Tribune* (March 3, 1991): C14.

8. Kotulak and Gorner, "Babies by Design," C14.

9. Elizabeth Rosenthal, "Technique for Early Prenatal Test Comes Under Question in Studies," *New York Times* (July 10, 1991).

10. William Booth, "Genetic Screening for Cystic Fibrosis Provokes Anxious Debate," *Washington Post* (August 10, 1991).

11. Booth, "Genetic Screening for Cystic Fibrosis."

12. Gina Kolata, "Fetal Sex Test Used as Step to Abortion," *New York Times* (December 25, 1988): A1.

13. Dorothy C. Wertz and John Fletcher, "Fatal Knowledge? Prenatal Diagnosis and Sex Selection," *Hastings Center Report* (May/June 1989): 21.

14. Dorothy C. Wertz and John Fletcher, "Sex Selection in India," *Hastings Center Report* (May/June 1989): 25; Jo McGowan, "In India, They Abort Females," *Newsweek* (January 30, 1989): 12.

15. Geoffrey Cowley, "Made to Order Babies," *Newsweek* (Winter/Spring 1990): 98.

16. Kotulak and Gorner, "Babies by Design," 14.

17. Cowley, "Made to Order Babies," 98.

18. Constance Holden, "On the Trail of Genes for IQ," *Science* 253: 1352.

19. Cowley, "Made to Order Babies," 94.

20. Booth, "Genetic Screening for Cystic Fibrosis."

21. Cowley, "Made to Order Babies," 98.

22. Cowley, "Made to Order Babies," 94.

23. Kotulak and Gorner, "Babies by Design," 14.

24. Gregg Levoy, "Wrongful Life, People Now Sue Just for Being Born," *San Francisco Chronicle* (March 25, 1990): 12.

25. Berman v. Allan, 404 A. 2d S 80 N.J. 421 (1979).

26. Berman v. Allan, 10.

27. Berman v. Allan, 11, 13.

28. Berman v. Allan, 12.

29. Berman v. Allan, 12.

30. Berman v. Allan, 12.

31. California: Cal. Civ. Code 43.6 (1991); Indiana: Ind. Code. Ann. 34-1-1-11 (1990); Minnesota: Minn. Stat. 145.424 (1990); Missouri: 188.130 R.S. Mo. (1989); South Dakota: S.D. Codified Laws 21-55-(1-4) (1991); Utah: Utah Code 78-11-(23-25) 1983).

32. See Turpin v. Sortini, 643 P. 2d 954 (1982); Procanik v. Cillo, 478 A. 2d. 755 (1984); Harbeson v. Parke-Davis, Inc., 656 P.2d 483 (1983).

33. Berman v. Allan, 14.

34. Berman v. Allan, 14.

35. Levoy, "Wrongful Life," 12.

36. Missouri: 188.130 R.S. Mo. (1989).

37. Missouri: 188.130 R.S. Mo. (1989).

38. Missouri: 188.130 R.S. Mo. (1989).

39. Missouri: 188.130 R.S. Mo. (1989).

CHAPTER 10
Designing Genes

1. Hans Jonas, "Contemporary Problems in Ethics from a Jewish Perspective," *Journal of Central Conference of American Rabbis* (January 1968): 36.

2. "Retailoring the Tailor," in *Encyclopedia Britannica, Book of the Year (1976),* Special Supplement, iv.

3. Jeremy Rifkin, *Algeny* (New York: Penguin, 1984), 11.

4. Quoted in Ted Howard and Jeremy Rifkin, *Who Should Play God? The Artificial Creation of Life and What It Means for the Future of the Human Race* (New York: Dell, 1977), 16.

5. Howard and Rifkin, *Who Should Play God?* 16.

6. Nicholas Wade, *The Ultimate Experiment, Man Made Evolution, DNA* (New York: Walker and Co., 1977), 20, 21; Jeremy Cherfas, *Man Made Life* (Oxford: Blackwell, 1982), 26–60; Edward Yoxen, *The Gene Business, Who Should Control Biotechnology?* (New York: Harper & Row, 1983), 71, 72.

7. Wade, *The Ultimate Experiment*, 26, 27.

8. Wade, *The Ultimate Experiment*, 26; Cherfas, *Man Made Life*, 85.

9. W. Gilbert, "The Minds Behind the Top 100: Who They Are, What They Think, What the Future Holds," *Science Digest* 93 (1985): 64, 65.

10. R. J. Kalter, "The New Biotech Agriculture, Unforeseen Economic Consequences," *Issues in Science and Technology* 1 (1985): 125–33.

11. U.S. Congress, Committee on Government Operations, *Designing Genetic Information Policy: The Need for an Independent Policy Review of Ethical, Legal, and Social Implications of the Human Genome Project* (Washington, D.C.: U.S. Government Printing Office, 1992).

12. Edmund L. Andrews, "U.S. Seeks Patent on Genetic Codes, Setting Off Furor," *New York Times* (October 21, 1991): A1, A12.

CHAPTER 11

A Discriminating Drug

1. C. S. Lewis, *The Abolition of Man* (New York: Macmillan, 1955), 71.

2. *Sports Illustrated* (July 8, 1991).

3. Lyle Alzado, as told to Shelly Smith, "I'm Sick and I'm Scared," *Sports Illustrated* (July 8, 1991): 21.

4. Alzado, "I'm Sick," 21.

5. Alzado, "I'm Sick," 21.

6. Alzado, "I'm Sick," 22, 23.

7. Alzado, "I'm Sick," 22, 23.

8. Joanne M. Schrof, "Pumped Up," *U.S. News & World Report* (June 1, 1992): 55.

9. Barry Werth, "How Short Is Too Short, Marketing Human Growth Hormone," *New York Times Magazine* (June 16, 1991): 14.

10. Werth, "How Short Is Too Short."

11. Werth, "How Short Is Too Short," 15.

12. Werth, "How Short Is Too Short," 28, 29.

13. Adult patient Consent Form, Study Number 91-CH-46, page 5, paragraph 9 (Bethesda, MD: The National Institutes of Health).

14. Consent Form, paragraph 8.

15. John Lantos, M.D., Mark Siegler, M.D., and Leona Cuttler, M.D., "Ethical Issues in Growth Hormone Therapy," *Journal of the American Medical Association* 261, no. 7 (February 17, 1989): 1021–1022.

16. Shaw Watanabe, Yukiko Tsunamatsu, Juichiro Fujimoto, and Atsushi Komiyama, "Leukemia in Patients Treated with Growth Hormone," *The Lancet* (May 21, 1988): 1159.

17. "Growth Hormone Treatment and Leukemia," *CMAJ* 139 (November 1, 1988).

18. Lawson Wilkins Pediatric Endocrine Society and the Human Growth Foundation of the United States, *Report on the International Workshop on Growth Hormone and Leukemia* (May 5, 1988): 1.

19. Lawson Wilkins Pediatric Endocrine Society, *Report*, 6.

20. Food and Drug Administration Response, June 21, 1991, response to Petition dated February 6, 1990, filed by the Foundation on Economic Trends (FET), Docket No. 90P-0075/CP, 3–4. (On file with FET, Washington, D.C.)

21. Food and Drug Administration Response to FET petition, 7.

22. Sally Lehrman, "The Fountain of Youth?" *Harvard Health Letter* 17, no. 8 (June 1992): 1–3.

23. Edward Yoxen, *The Gene Business, Who Should Control Biotechnology?* (New York: Harper & Row, 1983), 100.

24. Yoxen, *The Gene Business*, 99–100.

25. Martin Benjamin, James Muyskens, and Paul Saenger, "Short Children, Anxious Parents: Is Growth Hormone the Answer?" *The Hastings Center Report* (April 1984).

26. Werth, "How Short Is Too Short," 17.

27. Werth, "How Short Is Too Short," 47.

28. Werth, "How Short Is Too Short," 15.

29. Memorandum to Gilman Grave, M.D., Chair, ICRS, NICHD, Clinical Research Project Number 91-CH-46, "A Randomized, Double Blind, Placebo-Controlled Clinical Trial of the Effects of Growth Hormone Therapy on the Adult Height of Non Growth Hormone Deficient Children with Short Stature," November 27, 1990.

30. Agreement letter from Lilly Research Laboratories to Gordon B. Cutler, Jr., November 11, 1987.

31. Letter from Lilly Research Laboratories to Gordon B. Cutler, Jr., M.D., November 11, 1987.

32. Letter from Dr. Bernadine Healy, Director of NIH, to Andrew Kimbrell, counsel, Foundation on Economic Trends, May 3, 1991.

33. "NIH Hormone Tests with Children Draws Criticism of Group," *Wall Street Journal* (Eastern edition) (25 June 1992): A1.

34. Grave Memorandum; Minor Patient's Assent to Participate in a Clinical Research Study, this document is the minor consent form for project #91-CH-46; Richard Stone, "NIH to Size Up Growth Hormone Trials," *Science* 257 (August 7, 1992): 738.

35. Benjamin, Muyskens, and Saenger, "Short Children, Anxious Parents," 8.

36. Benjamin, Muyskens, and Saenger, "Short Children, Anxious Parents," 8.

37. Lantos, Siegler, and Cuttler, "Ethical Issues in Growth Hormone Therapy," 1022.

38. Werth, "How Short Is Too Short," 15.

39. Sec. 1904. Amendment of the Food, Drug and Cosmetic Act, November 29, 1990.

40. Werth, "How Short Is Too Short," 28.

41. Werth, "How Short Is Too Short," 28.

42. Lehrman, "The Fountain of Youth?" 1–3.

CHAPTER 12
Engineering Ourselves

1. James J. Nagle, "Genetic Engineering," *Bulletin of the Atomic Scientists* (December 1971): 44.

2. Christine Russell, "Protesters Highlight Genetics Meeting," *Washington Star* (March 8, 1977); personal interview with Jeremy Rifkin, October 15, 1991.

3. Marshall Nirenberg, "Will Society Be Prepared?" *Science* 633 (1967): 157

4. Cited in Jeremy Rifkin and Ted Howard, *Who Should Play God? The Artificial Creation of Life and What It Means for the Future of the Human Race* (New York: Dell, 1977), 155.

5. Leon R. Kass, "Making Babies—the New Biology and the 'Old' Morality," *The Public Interest* (Winter 1972): 53–54.

6. Editorial, *New York Times* (July 22, 1982).

7. *New York Times* (July December 29, 1982).

8. Paul Ramsey, "Genetic Therapy, a Theologian's Response," in M. Hamilton, ed., *The New Genetics and the Future of Man* (Grand Rapids, MI: Eerdmans, 1972), 163.

9. John C. Fletcher, "Evolution of Ethical Debate about Human Gene Therapy," *Human Gene Therapy* 1:55–68 (1990): 59.

10. *President's Commission for the Study of Ethical Problems in Medicine and Biomedical and Behavioral Research* (Washington D.C.: Government Printing Office, 1982), 2.

11. *President's Commission for the Study of Ethical Problems*, 48.

12. "The Rules for Reshaping Life," *New York Times* (December 29, 1982): editorial page.

13. "Resolution to Express the Conviction That Engineering Specific Traits into the Human Germline Not Be Attempted," distributed by the Foundation on Economic Trends, June 1983; see also, Philip J. Hilts, "Clergymen Ask Ban on Efforts to Alter Genes," *Washington Post* (June 8, 1983): 1.

14. "Biotechnology: Its Challenges to the Churches and the World" (World Council of Churches, Subunit on Church and Society, August 1989), 14.

15. "Biotechnology: Its Challenges to the Churches and the World," 14.

16. Foundation on Economic Trends, press release (January 30, 1989), 2.

17. Foundation on Economic Trends, press release (January 30, 1989), 2.

18. Department of Human Services, Public Health Service, National Institutes of Health, Recombinant DNA Advisory Committee, minutes of meeting, January 30, 1989, 31.

19. Foundation on Economic Trends, et. al. v. Louis D. Sullivan, M.D., United States District Court for the District of Columbia, Case No. 89-0222, Agreement of Dismissal, May 16, 1989; also see, Dan Sperling, "Genetic Research: Public Wins a Voice," *USA Today* (May 17, 1989): 1.

20. Robin Marantz Henig, "Dr. Anderson's Gene Machine," *New York Times Magazine* (March 31, 1991).

21. Henig, "Dr. Anderson's Gene Machine."

22. W. French Anderson, "Human Gene Therapy," *Science* 256 (May 8, 1992): 812 and fn. 34.

23. Peter Gorner and Ronald Kotulak, "Scientists Criticize Human Gene Therapy," *Chicago Tribune* (September 20, 1990): 1.

24. Gorner and Kotulak, "Scientists Criticize Human Gene Therapy," 2.

25. Sally Lehrman, "Breaking the Code," *Image, San Francisco Examiner* (May 19, 1991).

26. Larry Thompson, "The Ultimate Body Shop," *Warfield's, The Baltimore Business Monthly* (August 1989): 56.

27. Press Release from Genetic Therapy, Inc., "Genetic Therapy, Inc., Reports 1991 Financial Results," February 27, 1992, 1; see also, Genetic Therapy, Inc., Third Quarter 1991, Letter to Shareholders.

28. Anderson, "Human Gene Therapy," 808.

CHAPTER 13
The Beast Machines

1. Hugo Davenport, "The Brave New World of Animal Farm," *The Daily Telegraph* (London) (March 2, 1988): 1.

2. Keith Schneider, "Science Debates Using Tools to Redesign Life," *New York Times* (June 8, 1987): A17.

3. U.S. Congress, Office of Technology Assessment, *New Developments in Biotechnology: Patenting Life—Special Report, OTA-BA-370* (Washington, D.C.: U.S. Government Printing Office, April 1989), 100.

4. "Genetic Engineering?" *Agscene* no. 103 (Summer 1991): 22.

5. "Genetic Engineering?" 22.

6. "Genetic Engineering?" 22.

7. "Super Fish," *Equinox* (March/April 1987).

8. Mike Toner, "Cultivating 'Designer' Fish," *Atlanta Journal* (May 21, 1991).

9. Gregory Byrne, "A Fly-By-Light Discovery," *The Scientist* 1, no. 3 (December 15, 1986): 25.

10. OTA, *Patenting Life,* 13.

11. Reuters News Agency, "Genetic Juggling Raises Concerns," *Washington Times* (March 30, 1988).

12. Associated Press, "Animal Genetic Engineering Proves Controversial," *The Sioux City Journal* (November 10, 1987): A5.

13. Associated Press, "Animal Genetic Engineering Proves Controversial," A5.

14. Reuters News Agency, "Genetic Juggling Raises Concerns."

15. Schneider, "Science Debates Using Tools to Redesign Life," A17.

16. Schneider, "Science Debates Using Tools to Redesign Life," A17.

17. Schneider, "Science Debates Using Tools to Redesign Life," A17.

18. Marilyn Chase, "AIDS Virus Planted in Mouse Genes; Some Say Danger Offsets Research Gains," *Wall Street Journal* (January 13, 1988).

19. Chase, "AIDS Virus Planted in Mouse Genes."

20. Michael Specter, "Mice Develop AIDS-Like Symptoms," *Washington Post* (June 15, 1988): A14.

21. Chase, "AIDS Virus Planted in Mouse Genes."

22. Susan Okie, "NIH Scientists Introduce Genetic Code of AIDS Virus into Mice," *Washington Post* (December 6, 1987): A3.

23. Press Release, Foundation on Economic Trends, December 15, 1987.

24. Chase, "AIDS Virus Planted in Mouse Genes," X.

25. Okie, "NIH Scientists Introduce Genetic Code of AIDS Virus Into Mice," A3.

26. Chase, "AIDS Virus Planted in Mouse Genes."

27. Chase, "AIDS Virus Planted in Mouse Genes."

28. Michael Specter, "Lab Mishap Destroys AIDS Mice," *Washington Post* (December 8, 1988): A3.

29. Paolo Lusso, Fulvia Di Marco Veronese, Barbara Ensoli, Genoveffa Franchini, Christine Jemma, Susan E. DeRocco, V. S. Kalyanaraman, and Robert C. Gallo, "Expanding HIV-1 Cellular Tropism by Phenotypic Mixing with Murine Endogenous Retroviruses," *Science* 247 (February 16, 1990): 848.

30. Jean Marx, "Concerns Raised About Mouse Models for AIDS," *Science* 247 (February 16, 1990): 809.

31. Lusso, et al., "Expanding HIV-1 Cellular Tropism," 851.

32. Anne Simon Moffat, "Transgenic Animals May Be Down on the Pharm," *Science* (October 4, 1991).

33. Moffat, "Transgenic Animals."

34. Randi Hutter Epstein, "Beef Sales Up Despite 'Mad Cow' Disease," *Los Angeles Times* (October 13, 1991): A25.

35. Keith Schneider, "AIDS-like Cow Virus Found at Unexpectedly High Rate," *New York Times* (June 1, 1991): A1.

CHAPTER 14
The Patenting of Life

1. Ted Howard, "The Case Against Patenting Life," brief on Behalf of the People's Business Commission, Amicus Curiae, in the Supreme Court of the United States, No. 79-136, 29.

2. Quoted in Virginia Morris, "Human Genes Not So Special," *New Haven Register* (August 28, 1988): B1.

3. U.S. Congress, Office of Technology Assessment, *New Developments in Biotechnology: Patenting Life—Special Report, OTA-BA-370* (Washington, D.C.: U.S. Government Printing Office, April 1989), 37.

4. OTA, *Patenting Life*, 37–38.

5. "New Patents Challenge Congress and Courts," *Scientific American* (September 1988): 128.

6. Sharon McAuliffe and Kathleen McAuliffe, *Life for Sale* (New York: Coward, McCann & Geoghegan, 1981), 203–4.

7. Christopher Anderson, "U.S. Patent Application Stirs Up Gene Hunters," *Nature* 353 (October 10, 1991), 485.

8. Congressional Record, Senate, April 2, 1992, S 4722, citing Leslie Roberts, "NIH Gene Patents, Round Two," *Science* (February 21, 1992).

9. Robin Herman, "NIH Genes Researcher Is Leaving for His Own Lab," *Washington Post* (July 7, 1992): Health, 4.

10. Edmund L. Andrews, "U.S. Seeks Patent on Genetic Codes, Setting Off Furor," *New York Times* (October 21, 1991): A1, A12.

11. Michael Waldholz, Hilary Stout, "A New Debate Rages Over the Patenting of Gene Discoveries," *Wall Street Journal* (April 17, 1992): A1, A6.

12. Susan Watts, "A Matter of Life and Patents," *New Scientist* (January 12, 1991): 57.

13. Watts, "A Matter of Life and Patents," 47.

14. Steve Connor, "Breasts Provoke Patent Conflict," *The Independent* (London) (February 19, 1992).

15. Giovanna Brel, "An Illinois Biochemist Wins a Crucial Patent Fight, and a New Era of Life in a Test Tube Begins," *People* (July 14, 1980): 38.

16. Sidney A. Diamond, Commissioner of Patents and Trademarks, petitioner, v. Ananda M. Chakrabarty et al., 65 L ed 2d 144, June 16, 1980, 144–47.

17. Diamond v. Chakrabarty, 148.

18. See In re Bergy, 563 F. 2d 1031 (1975).

19. Parke, Acting Commissioner of Patents and Trademarks v. Flook, 437 U.S. 584, 596 (1978).

20. Parke v. Flook.

21. Diamond v. Chakrabarty, 148.

22. McAuliffe and McAuliffe, *Life for Sale*, 199.

23. Howard, "The Case Against Patenting Life," 29.

24. Diamond v. Chakrabarty, 152.

25. Diamond v. Chakrabarty, 158.

26. Diamond v. Chakrabarty, 155.

27. Diamond v. Chakrabarty, 154.

28. Brel, "An Illinois Biochemist Wins a Crucial Patent Fight," 37.

29. OTA, *Patenting Life*, 55.

30. McAuliffe and McAuliffe, *Life for Sale*, 205.

31. McAuliffe and McAuliffe, *Life for Sale*, 205.

32. McAuliffe and McAuliffe, *Life for Sale*, 205.

33. McAuliffe and McAuliffe, *Life for Sale*, 204.

34. Leon R. Kass, "Patenting Life," *Commentary* (December 1981), 56.

35. United States Patent, Leder et al., Patent Number:4, 736, 866, Date of Patent April 12, 1988, 1–3; OTA, *Patenting Life*, 99; Malcolm Gladwell, "Harvard Scientists Win Patent for Genetically Altered Mouse," *Washington Post* (April 12, 1988): A1.

36. United States Patent, Leder et al., 1–3; *Patenting Life*, 99; Gladwell, "Harvard Scientists Win Patent for Genetically Altered Mouse," A1.

37. OTA, *Patenting Life*, 99.

38. See DuPont advertisement "OncoMouse Transgenic Animal," in files of author (also available from DuPont, 1-800-551-2121).

39. David Brown, "Little Progress in War on Breast Cancer," *Washington Post* (December 12, 1991): A8.

40. U.S. Patent and Trademark Office, Official Gazette, December 29, 1992, p. 2924; "U.S. Patents are granted for three laboratory mice," *The Washington Post* (Associated Press), 30, December, 1992.

41. Testimony of Michael Glough, Office of Technology Assessment, U.S. Congress, Before the Subcommittee on Intellectual Property and Judicial Administration House Committee on the Judiciary, November 20, 1991, on "Patents and Biotechnology," 3.

42. OTA, *Patenting Life*, 74–75.

43. U.S. Patent and Trademark Office, *Animals–Patentability* (Washington, D.C.: U.S. Patent and Trademark Office, April, 7, 1987).

44. PTO ruling *Animals—Patentability*.

45. See, for example, "As Congress Sleeps, Here Come the Geeps," *Philadelphia Inquirer* (February 21, 1988): editorial page; "Patenting Life: New Federal Policy Is an Invitation to Trouble," *Tucson Daily Star* (April 20, 1987): editorial page.

46. Keith Schneider, "New Animal Forms Will Be Patented," *New York Times* (April 17, 1987).

47. Introduction of Moratorium on Animal Patenting, Statement of Senator Mark Hatfield, Congressional record - Senate, S 7268, May 28, 1987.

48. Sally Lehrman, "Biotech Firms Persuaded Bush to Reject Earth Treaty" (June 10, 1992): A-1.

49. Kass, "Patenting Life," 56.

50. Statement of Religious Leaders Against Animal Patenting to the Committee on the Judiciary Subcommittee on Courts, Civil Liberties, and the Administration of Justice, U.S. House of Representatives, July 22, 1987.

51. "Life Industrialized," *New York Times* (February, 22 1988), editorial page.

CHAPTER 15

A Monopoly on Humanity

1. Moore v. Regents of the University of California, 202 Cal App., 1270.

2. Beverly Merz, "Whose Cells Are They, Anyway," *American Medical News* (March 23/30, 1990): 7, 8.

3. Merz, "Whose Cells Are They, Anyway," 7, 8.

4. U.S. Congress, Office of Technology Assessment, *New Developments in Biotechnology: Ownership of Human Tissues and Cells—Special Report,* OTA-BA-337 (Washington, D.C.: U.S. Government Printing Office, March 1987), 25.

5. OTA, *Ownership of Human Tissues and Cells,* 26.

6. Opinion, Moore v. The Regents of the University of California et al., Supreme Court of the State of California, file # S006987, slip op. 2, 3; William Carlsen, "Key Ruling by State Court on Body Cells," *San Francisco Chronicle* (July 10, 1990): 1.

7. Opinion, Moore v. The Regents of the University of California et al., Supreme Court of the State of California, file # S006987, slip op. 2, 3; Carlsen, "Key Ruling by State Court on Body Cells," 1.

8. Opinion, Moore v. The Regents of the University of California et al., Supreme Court of the State of California, file # S006987, slip op. 2, 3; Carlsen, "Key Ruling by State Court on Body Cells," 1.

9. Carlsen, "Key Ruling by State Court on Body Cells," 1; Moore v. Regents of the University of California, 202 Cal App. 3d 1230, 1239-40. July 1988.

10. Moore v. The Regents of the University of California et al., Supreme Court of the State of California, 8, 9.

11. Moore v. Regents of the University of California, 202 Cal App., 1248.

12. Moore v. Regents of the University of California, 202 Cal App., 1254.

13. Moore v. Regents of the University of California, 202 Cal App., 1249.

14. Moore v. Regents of the University of California, 202 Cal App., 1270.

15. Moore v. The Regents of the University of California et al., Supreme Court of the State of California, 16.

16. Moore v. The Regents of the University of California et al., Supreme Court of the State of California, 23.

17. See United States Patent, # 5,061,620, Oct. 29, 1991, "Human Hematopoietic Stem Cell."

18. Michael Waldholz and Hilary Stout, "A New Debate Rages Over the Patenting of Gene Discoveries," *Wall Street Journal* (April 17, 1992): A6.

19. Sharon Brownlee, "Staking Claims on the Human Body," *U.S. News & World Report* (November 18, 1991): 89.

CHAPTER 16

A Clone Just for You

1. Rebecca Kolberg, "Human Embryo Cloning Reported," *Science* (October 29, 1993): 652; Philip Elmer-Dewitt, "Cloning: Where Do We Draw the Line?" *Time* (November 8, 1993): 65.

2. "Human Embryo Cloning Reported," 652; "Cloning: Where Do We Draw the Line?" 65; Gina Kolata, "Scientist Clones Human Embryos, and Creates an Ethical Challenge," *New York Times* (October 24, 1993): 1.

3. "Human Embryo Cloning Reported," 652; "Cloning: Where Do We Draw the Line?" 65; "Scientist Clones Human Embryos, and Creates an Ethical Challenge," 1.

4. D. S. Halacy, Jr., *Genetic Revolution, Shaping Life for Tomorrow* (New York: Harper & Row, 1974), 160.

5. Halacy, *Genetic Revolution,* 159–60.

6. Halacy, *Genetic Revolution,* 159–60.

7. Halacy, *Genetic Revolution,* 161.

8. Halacy, *Genetic Revolution,* 161.

9. Halacy, *Genetic Revolution,* 162.

10. *Holstein World,* special Dairy Expo edition (June 1990): 58–59.

11. Keith Schneider, "Better Farm Animals Duplicated by Cloning," *New York Times* (February 17, 1988): A1.

12. Jean L. Marx, "Cloning Sheep and Cattle Embryos," *Science* 239 (January 28, 1988): 463–64; N. L. First, "New Animal Breeding Techniques and Their Application," *Journals of Reproduction & Fertility* (Supplement, 1990): 6.

13. First, "New Animal Breeding Techniques and Their Application," 6.

14. Claudia Deutsch, "An Old Pol Tackles a New Post," *New York Times* (May 22, 1988): 1.

15. Letter from Frank J. Hartdegan, Ph.D., Research Associate, W. R. Grace, to Mr. Robert W. Ericson, Director, Research Administration, Financial, University of Wisconsin, Madison, October 1, 1986.

16. Letter from Vincent F. Simon, Vice President, Research Division W. R. Grace, to Neal L. First, Ph.D, Professor, Department of Meat and Animal Science, University of Wisconsin, October 6, 1986.

17. Joel McNair, "ABS Receives Embryo Cloning Patent," *Agriview* (May 2, 1991).

18. Mara Bovsun, "Calf Clones Come to Market Despite Mysterious Problems," *Biotechnology Newswatch* (July 6, 1992): 3.

19. Neville Hodgkinson, "Lab Trial Spawns 'Freak' Animals," *The Sunday Times (London)* (March 1992): 1.

20. Bovsun, "Calf Clones," 1.

21. Bovsun, "Calf Clones," 1, 3.

22. Hodgkinson, "Lab Trial Spawns 'Freak' Animals," 1.

23. Bovsun, "Calf Clones," 1.

24. Mike Flaherty, "Giant Calves Slow Cloning," *Wisconsin State Journal* (July 8, 1992): A1.

25. Hodgkinson, "Lab Trial Spawns 'Freak' Animals," 1.

26. Kolberg, "Human Embryo Cloning Reported," 652.

27. "Human Embryo Cloning Reported," 652.

28. Jeremy Cherfas, *Man Made Life* (Oxford: Blackwell, 1982), 231.

29. Cherfas, *Man Made Life*, 231.

30. Edward Yoxen, *The Gene Business, Who Should Control Biotechnology?* (New York: Harper & Row, 1983), 115.

31. Cherfas, *Man Made Life*, 231–32.

32. Halacy, *Genetic Revolution*, 163.

33. Warren E. Leary, "Scientists Seek Lincoln DNA to Clone for a Medical Study," *New York Times* (February 10, 1991): A1.

34. Bob Dart, "Bid to Clone Lincoln DNA Gets Support," *Atlanta Journal/The Atlanta Constitution* (May 3, 1991).

35. Jeremy Rifkin and Ted Howard, *Who Should Play God?* (Laurel Books, 1980), 125.

CHAPTER 17

Crawling Machines and the Invisible Hand

1. Quoted in Tony Augarde, ed., *The Oxford Dictionary of Modern Quotations* (Oxford, New York: Oxford University Press, 1991), 190.

2. G. K. Chesterton, *The Father Brown Omnibus, The Purple Wig* (New York: Dodd, Mead & Company, 1951), 222.

3. "Trading Flesh Around the Globe," *Time* 137 (June 17, 1991): 61.

4. Pakrash Chandra, "Kidneys for Sale," *World Press Review* 38 (February 1991): 53.

5. Georgina Ferry, "New Cells for Old Brains," *New Scientist* (March 24, 1988): 54–57.

6. Gina Kolata, "More Babies Being Born to Be Donors of Tissues," *New York Times* (June 4, 1991): C3.

7. See Sandra Blakeslee, "Human Tissues in Lab Mice," *New York Times* (October 30, 1990): A1.

8. Rochelle Sharpe, "Suit Challenges Constitutionality of Surrogate Parenting," *Michigan State Journal* (October 3, 1986): 1B.

9. Neville Hodgkinson, "Lab Trial Spawns 'Freak' Animals," *The Sunday Times (London)* (March 1992): 1.

10. Marilyn Chase, "AIDS Virus Planted in Mouse Genes," *Wall Street Journal* (January 13, 1988).

11. Edmund L. Andrews, "U.S. Seeks Patent on Genetic Codes," *New York Times* (October 21, 1991): A1, A12.

12. Steve Connor, "Breasts Provoke Patent Conflict," *The Independent (London)* (February 19, 1992).

CHAPTER 18

The Body as Technology

1. Quoted in Anson Rabinbach, *Monodology* (Berkeley, Los Angeles: University of California Press, 1990), sec. 64.

2. G. K. Chesterton, "What's Wrong with the World," in G. K. Chesterton, *Collected Works* (San Francisco: Ignatius Press, 1987), 61.

3. Virginia Morris, "Human Genes Not So Special," *New Haven Register* (August 28, 1988): 1.

4. "Life Industrialized," *New York Times* (February 22, 1988): editorial page.

5. Quoted in David F. Channel, *The Vital Machine* (New York, Oxford: Oxford University Press, 1991), 125.

6. Maurice A. Finocchiaro, *The Galileo Affair, A Documentary History* (Berkeley: University of California Press, 1989), 292.

7. Mumford, *The Myth of the Machine*, 57, 58.

8. Mumford, *The Myth of the Machine*, 58.

9. Scott Buchanan, *So Reason Can Rule* (New York: Farrar, Strauss, Giroux, 1982), 296.

10. Channel, *The Vital Machine*, 22.

11. Jeremy Rifkin, *Time Wars* (New York: Henry Holt, 1987), 176.

12. William Barrett, *The Death of the Soul* (Garden City, NY: Doubleday, 1987), 10.

13. Quoted in Rifkin, *Time Wars*, 175.

14. Aram Vartanian, *Diderot and Descartes* (Westport, CT: Greenwood Press, 1953), 213.

15. Leonora Cohen Rosenfield, *From Beast-Machine to Man-Machine*, (New York: Octagon Books, 1968), 136, 137.

16. Rosenfield, *From Beast-Machine to Man-Machine*, 99.

17. Rosenfield, *From Beast-Machine to Man-Machine*, 8.

18. Rosenfield, *From Beast-Machine to Man-Machine*, 54.

19. Quoted in Rosenfield, *From Beast-Machine to Man-Machine*, 224, n.90.

20. Quoted in Anson Rabinbach, *The Human Motor* (Berkeley, Los Angeles: University of California Press, 1990), 64.

21. Channel, *The Vital Machine*, 40.

22. Quoted in Vartanian, *Diderot and Descartes*, 205, 206.

23. Channel, *The Vital Machine*, 44.

24. Donald Worster, *Nature's Economy: The Roots of Ecology* (Garden City, NY: Anchor Books, 1979), 40.

CHAPTER 19
The Human Motor

1. Quoted in Lewis Mumford, *The Myth of the Machine, The Pentagon of Power* (New York: Harcourt Brace Jovanovich, 1970), 56.

2. Owen Barfield, *Poetic Diction, A Study in Meaning* (Middletown, CT: Wesleyan University Press, 1973), 22.

3. Anson Rabinbach, *The Human Motor* (Berkeley, Los Angeles: University of California Press, 1990), 2.

4. Rabinbach, *The Human Motor*, 67.

5. Richard Weaver, *The Ethics of Rhetoric* (South Bend, IN: Regnery/Gateway, 1953), 217, 218.

6. Rabinbach, *The Human Motor*, 68.

7. Rabinbach, *The Human Motor*, 74.

8. Jeremy Rifkin, *Time Wars* (New York: Henry Holt, 1987), 106, 107.

9. Committee on Government Operations, *Designing Genetic Information Policy: The Need for an Independent Policy Review of the Ethical, Legal, and Social Implications of the Human Genome Project*, sixteenth report (Washington, D.C.: U.S. Government Printing Office, 1992), 19–21.

10. U.S. Congress, Office of Technology Assessment (OTA), *Genetic Monitoring and Screening in the Workplace*, OTA-BA-445 (Washington, D.C.: U.S. Government Printing Office, October 1990), 177–78.

11. See Committee on Government Operations, *Designing Genetic Information Policy*, 15–21.

12. Judith Rodin, Ph.D., "Body Mania," *Psychology Today* (January/February 1992): 55.

13. Rodin, "Body Mania," 55, 56.

14. Robert T. Grieves, "Muscle Madness," *Forbes* (May 30, 1988): 216.

15. Rodin, "Body Mania," 58.

16. Rodin, "Body Mania," 58.

17. Mark Haller, *Eugenics: Hereditarian Attitudes in American Thought* (New Brunswick, NJ: Rutgers University Press, 1963), 3.

18. Quoted in Haller, *Eugenics*, 76.

19. Quoted in Ted Howard and Jeremy Rifkin, *Who Should Play God?* (New York: Dell, 1977), 52.

20. Howard and Rifkin, *Who Should Play God?*, 74, 75.

21. Larry Azar, *Philosophy and Ideology* (Des Moines, IA: Kendall Hunt), 47.

22. Daniel J. Kevles, *In the Name of Eugenics* (New York: Alfred A. Knopf, 1985), ix.

23. Quoted in Howard and Rifkin, *Who Should Play God?*, 51.

24. Haller, *Eugenics*, 47.

25. Kevles, *In the Name of Eugenics*, 12.

26. Haller, *Eugenics*, 19.

27. Haller, *Eugenics*, 17.

28. Howard and Rifkin, *Who Should Play God?*, 54.

29. Haller, *Eugenics*, 23.

30. George J. Marlin and Richard P. Rabatin, "G. K. Chesterton and Eugenics," *Fidelity* (June 1990): 33.

31. Quoted in Haller, *Eugenics*, 83.

32. Quoted in Haller, *Eugenics*, 41.

33. Quoted in Marlin and Rabatin, "G. K. Chesterton and Eugenics," 33.

34. Quoted in Robert N. Proctor, *Racial Hygiene, Medicine Under the Nazis* (Cambridge, MA: Harvard University Press, 1988), 100.

35. Proctor, *Racial Hygiene*.

36. Kevles, *In the Name of Eugenics*, 166.

37. Kevles, *In the Name of Eugenics*, 166.

38. Skinner v. Oklahoma, 62 U.S. 110, 114 (1942).

39. "Crimes-in-Genes Conference Proves Too Risky for NIH Funds," *Biotechnology Newswatch* 12, no. 17 (September 7, 1992).

40. Stanley Peele, "Second Thoughts About Gene for Alcoholism," *Atlantic Monthly* (August 1990): 52–54.

CHAPTER 20
The Gospel of Greed

1. George Dalton, ed. and Introduction, *Primitive, Archaic and Modern Economies, Essays of Karl Polanyi* (Boston: Beacon Press, 1968), ix.

2. John Maynard Keynes, "Economic Possibilities for Our Grandchildren" (1930), in Keynes, *Essays in Persuasion* (New York: Norton, 1963), 372.

3. Robert H. Nelson, *Reaching for Heaven on Earth* (Savage, MD: Rowman & Littlefield, 1991), 2.

4. Nelson, *Reaching for Heaven on Earth*, 99.

5. Quoted in Nelson, *Reaching for Heaven on Earth*, 6.

6. Quoted in Nelson, *Reaching for Heaven on Earth*, 3.

7. Robert L. Heilbroner, *The Making of Economic Society*, 6th ed. (Englewood Cliffs, NJ: Prentice-Hall, 1980), 23.

8. Karl Polanyi, *The Great Transformation* (Boston: Beacon Press, 1944), 68.

9. Polanyi, *The Great Transformation*, 62.

10. Nelson, *Reaching for Heaven on Earth*, 97.

11. Quoted in Jeremy Rifkin, *Algeny* (New York: Viking Press, 1983), 93.

12. Adam Smith, *The Wealth of Nations*, in *Masterworks of Economics*, vol. 1 (New York: McGraw-Hill, 1973), 167.

13. Jeremy Rifkin, *Biosphere Politics* (New York: Crown, 1991), 173.

14. Polanyi, *The Great Transformation*, 135.

15. Nelson, *Reaching for Heaven on Earth*, 6.

16. Quoted in Nelson, *Reaching for Heaven on Earth*, 237.

17. Quoted in Nelson, *Reaching for Heaven on Earth*, 96.

18. Dalton, ed. *Primitive, Archaic and Modern Economies*, 61–62.

CHAPTER 21

Satanic Mills

1. Thomas Hobbes, *Leviathan*, part 2, edited by C. B. Macpherson (Harmondsworth, England: Penguin Books, 1979), 295.

2. Lewis Mumford, *Technics and Civilization* (New York and London: Harcourt Brace Jovanovich, 1963), 146.

3. Smith, *Wealth of Nations*, 60.

4. Jeremy Rifkin, *Time Wars* (New York: Henry Holt, 1987), 105.

5. Nelson, *Reaching for Heaven on Earth*, 99.

6. Quoted in Jeremy Rifkin, *Biosphere Politics* (New York: Crown, 1991), 41.

7. Robert L. Heilbroner, *The Making of Economic Society*, 6th ed. (Englewood Cliffs, NJ: Prentice-Hall, 1980), 82.

8. Heilbroner, *The Making of Economic Society*, 83.

9. Gina Kolata, "More Children Are Employed, Often Perilously," *New York Times* (June 21, 1992): A1.

10. Melissa Hackey, "Injuries and Illnesses in the Workplace, 1989," *Monthly Labor Review* (May 1991): 34, 35.

11. Quoted in Jeremy Rifkin, *The Green Lifestyle Handbook*, Andrew Kimbrell and Kirk B. Smith, eds. (New York: Henry Holt, 1990), xiii.

CHAPTER 22

At the Crossroad

1. Sharon and Kathleen McAuliffe, *Life for Sale* (New York: Coward, McCann & Geoghagen, 1981), 221.

2. Rorie Sherman, "Bioethics Debate: Americans Polled in Bioethics," *National Law Journal* (May 13, 1991), 1.

3. Sherman, "Bioethics Debate," 1.

4. Sherman, "Bioethics Debate," 1.

5. Sherman, "Bioethics Debate," 1.

6. "Fetal Research: Right or Wrong?" *Redbook* (December 1990), 170.

7. Sherman, "Bioethics Debate," 1.

8. Sherman, "Bioethics Debate," 1.

9. Dr. Thomas J. Hoban and Dr. Patricia A. Kendall, "Consumer Attitudes About the Use of Biotechnology in Agriculture and Food Production," Interim Report to the USDA, July 1992, 38.

10. Hoban and Kendall, "Consumer Attitudes," 38.

11. "Vast Majority of Americans Support Gene Therapy and Research, March of Dimes Survey Finds," news release, September 29, 1992, March of Dimes, White Plains, New York.

12. "Vast Majority of Americans Support Gene Therapy and Research, March of Dimes Survey Finds."

13. Darryl Macer, "Public Opinion on Gene Patents," *Nature* 358 (July 23, 1992).

CHAPTER 23

The Body Revolution

1. Quoted in Jeremy Rifkin, ed., *The Green Lifestyle Handbook* (New York: Holt, 1990), 88.

2. Quoted in Jeremy Rifkin, Andrew Kimbrell, Kirk Smith, eds., *The Green Lifestyle Handbook* (New York: Henry Holt, 1990), 1.

3. U.S. Congress, Office of Technology Assessment, *New Developments in Biotechnology, Ownership of Human Tissues and Cells—Special Report*, OTA-BA-337 (Washington, D.C.: U.S. Government Printing Office, March 1987), 143, 144.

4. Rupert Sheldrake, *The Presence of the Past; Morphic Resonance and the Habits of Nature* (Times Books, 1988), 76.

5. Sheldrake, *The Presence of the Past*, 76.

6. Quoted in Andrew Kimbrell, "Body Wars," *Utne Reader* (May/June 1992): 64.

7. Bronislaw Malinowski, *Argonauts of the Western Pacific* (New York: Dutton, 1922), 45.

8. Malinowski, *Argonauts*, 47.

9. Thomas Murray, "Gifts of the Body and the Needs of Strangers," *Hastings Center Report* (April 1987): 31.

10. Quoted in Jeremy Rifkin, *Biosphere Politics* (New York: Crown, 1991), 25, 26.

11. Murray, "Gifts of the Body," 31.

12. Murray, "Gifts of the Body," 31.

13. Lewis Hyde, *The Gift* (New York: Vintage, 1983), 22–23.

Index

Acknowledgments

THIS BOOK IS the result of several years of research, litigation, discussions, and debates in bioethics and biotechnology, work undertaken with many people whose views and actions inform almost every page. To thank them all as they deserve would require a book in itself. However I would like to especially thank Jeremy Rifkin for introducing me to this subject and for his friendship over many years. Special thanks are also due Nicole Kerber whose tireless research was an essential part of the writing of this book. I am also grateful to Sean Cunningham and my brother, Mark Kimbrell, for their indispensable work in sharpening both my prose and my ideas. Finally I would like to thank my wife, Kaiulani Lee, both for her patience with me during the writing of this book and for her many fine editorial suggestions.